MICROCOMPUTER BUSSES

MICROCOMPUTER BUSSES

R. M. Cram

University of California, San Diego
La Jolla, California
and
Thermo Electron Technologies
San Diego, California

ACADEMIC PRESS, INC.
Harcourt Brace Jovanovich, Publishers
San Diego   New York   Boston   London   Sydney   Tokyo   Toronto

This book is printed on acid-free paper. ∞

Copyright © 1991 by ACADEMIC PRESS, INC.
All Rights Reserved.
No part of this publication may be reproduced or transmitted in any form or by any means, electronic or mechanical, including photocopy, recording, or any information storage and retrieval system, without permission in writing from the publisher.

Academic Press, Inc.
San Diego, California 92101

United Kingdom Edition published by
Academic Press Limited
24–28 Oval Road, London NW1 7DX

Library of Congress Cataloging-in-Publication Data

Cram, R. M.
   Microcomputer busses / R.M. Cram.
     p. cm.
   Includes index.
   ISBN 0-12-196155-9
   1. Microcomputers --Buses. I. Title.
  TK7895.B87C73   1991
621369'16--dc20
                                   91-19191
                                       CIP

PRINTED IN THE UNITED STATES OF AMERICA
91 92 93 94   9 8 7 6 5 4 3 2 1

**Books are to be returned on or before
the last date below.**

0 3 JUN 1993

-8 DEC 1994

Contents

Preface   ix

1

Basic Bus Concepts   1
- 1.0 Bus Background   1
- 1.1 Bus Definition   2
- 1.2 Design Trade-Off Assessment Factors   2
- 1.3 Signal Groups and Uses   4
- 1.4 Transmission-Line Concepts   6

2

A Comparison of Busses   17
- 2.0 Limitations   17
- 2.1 A Basis for Comparison   18
- 2.2 Which Busses   19
- 2.3 Bus Comparison   19

3

Software/Hardware Interactions   27
- 3.0 What and Why, or "A Touch of Philosophy"   27
- 3.1 The Big Picture   28
- 3.2 Bus Interaction   32

4

MULTIBUS I 37
- 4.0 MULTIBUS I Overview 37
- 4.1 iSBX Bus 39
- 4.2 iLBX Bus 39
- 4.3 Multichannel I/O Bus 40
- 4.4 MULTIBUS System Bus 41
- 4.5 MULTIBUS Data Transfers 49
- 4.6 Multiple Masters 52
- 4.7 Interrupts 56

5

MULTIBUS II 59
- 5.0 MULTIBUS II Overview 59
- 5.1 Basic Features and Capabilities of MULTIBUS II 60
- 5.2 MULTIBUS II Mechanical Specifications 60
- 5.3 MULTIBUS II Operations 63
- 5.4 MULTIBUS II Line Descriptions 67
- 5.5 Timing Requirements for Bus Operations 73
- 5.6 Electrical Specifications 80

6

VMEbus 81
- 6.0 VMEbus Overview 81
- 6.1 Basic Features and Capabilities of the VMEbus 82
- 6.2 VMEbus Mechanical Specifications 82
- 6.3 VMEbus Cycles 86
- 6.4 VME Functional Modules 87
- 6.5 Data Transfer Bus 88
- 6.6 Data Transfer Arbitration Bus 101
- 6.7 Priority Interrupt Bus 108

Contents   vii

7

NuBus   115

- 7.0   NuBus Overview   115
- 7.1   NuBus Mechanical Specifications   116
- 7.2   Bus Lines   119
- 7.3   Types of Bus Cycles and Transactions   121
- 7.4   Bus Arbitration   128
- 7.5   Geographic Addressing   130
- 7.6   Utility Functions   130
- 7.7   NuBus Electrical Characteristics   132

8

PC/XT/AT Bus   135

- 8.0   PC/XT/AT Bus Overview   135
- 8.1   Basic Features and Capabilities of PC Busses   136
- 8.2   PC Bus Mechanical Specifications   137
- 8.3   Bus Lines   139
- 8.4   PC, XT, and AT Bus Cycles   146

9

STD Bus   153

- 9.0   STD Bus Overview   153
- 9.1   STD Bus Mechanical Outline   154
- 9.2   STD Bus Signal Lines   155
- 9.3   Data Transfer Operations   159
- 9.4   Signal Voltage Tolerances   163

10

Programmable Logic Devices   165

- 10.0   Programmable Logic Devices Overview   165
- 10.1   Boolean Equation Review   166

- 10.2 State Machine Review   168
- 10.3 PLD Architecture   171
- 10.4 Programming Programmable Logic Devices   182
- 10.5 Metastability   183
- 10.6 Programming Examples   184

11

Field-Programmable Gate Arrays   203

- 11.0 Field-Programmable Gate Arrays Overview   203
- 11.1 FPGA Speed Considerations   204
- 11.2 FPGA Architecture   204
- 11.3 Configuration   212
- 11.4 FPGA Development Process   212
- 11.5 Future Trends   214

12

MULTIBUS I Design Example   217

- 12.0 Summary   217
- 12.1 MULTIBUS Digital-to-Analog Converter Board   217

13

VMEbus Design Example   231

- 13.0 Summary   231
- 13.1 VMEbus Memory Board   231

Index   239

Preface

Microcomputer Busses is intended to be a tool for use by systems engineers or design engineers who either are in the process of designing a microcomputer-based system using microcomputer busses or are trying to develop a system architecture capable of performing a task for a set of requirements. Engineering students will find this book useful because it provides a good introduction to the basic features and capabilities of a range of popular commercial busses. As is the objective of any engineering text, this book prepares the engineer or engineering student to successfully use the material presented in the solution of engineering problems.

Any microcomputer or microcomputer bus is a tool by which engineering problems are solved. It is not an end in itself but only a tool providing a means to an end. There are a number of different types of bus tools in existence, and the suitability and cost of a bus solution for any given problem depends on the nature of the problem and the capabilities of the bus. One of the more difficult problems a design engineer must face is matching the cost and performance requirements of an engineering problem with the correct bus from the "bus toolbox." In general, more complex and newer busses have greater capability but also greater cost. Performance comes at a price and it is foolish to include unnecessary capabilities into a piece of hardware.

The first three chapters, which are particularly oriented toward the student, present in a readable format a tutorial on basic bus and transmission-line concepts, a comparison of busses, and a discussion of the process by which a program produces a given set of hardware responses or operations on a bus. Chapters 10 and 11, also student oriented, introduce the process of designing with programmable logic devices (PLDs) and gate arrays. Since the use of these devices has grown tremendously in the past few years, it is important for the student to have a good grasp of the capabilities of these devices. For a working engineer, these chapters may provide a new look at a familiar topic or an introduction to an area of design that may not have been previously used.

Chapters 4 and 9 introduce a number of computer busses. The busses covered are MULTIBUS I, MULTIBUS II, VMEbus, STD bus, NuBus, and PC/XT bus. The performance capabilities, electrical interface, and mechanical interface requirements for each of these busses are presented. All of these busses are being used extensively in the commercial marketplace and, in most cases, a national standard has been adopted by IEEE, forcing the bus product manufacturers to adhere to a common set of interface requirements. This results in products that are interchangeable and standardized.

This "main body" of the book is useful to the system architect and designer as well as to the student and technician. It provides a look into the details of the most important features in the operation of each bus. Included is information useful for predicting system performance, configuring a system, performing design functions, and troubleshooting. The intent of this book is not to provide the same level of detail present in the national specifications but instead to offer a simplified view of each bus and its most important characteristics. This book does not replace the national standards but is a supplement to those standards by providing a condensed (and abbreviated) view of each of the busses listed.

The last section of the book contains design examples of existing sample boards that have been commercially produced. These examples provide a means by which an engineer or student can examine a currently produced design to see the various methods by which the design process is implemented. This is useful to both the student and engineer in that a given interface to a bus can be implemented in a number of ways. Some ways may be considerably less efficient than others, but it is useful to see how a design has been implemented to make it possible.

I would like to thank the following companies who permitted me to reproduce proprietary data for inclusion in this book: Data I/O Corporation, PLX Corporation, Altera Corporation, Xilinx Corporation, Datel Corporation, Actel Corporation, and Dy-4 Systems Inc.

Basic Bus Concepts

1.0 Bus Background

On the surface, the design and use of any computer bus is not a profound topic. But it is an important topic for any system architect or engineering manager attempting to make an intelligent decision regarding the type and suitability of a given bus or for a design engineer trying to produce a functional design or make a reliable, cost-effective system for a given application. Busses are so fundamental to all computer systems that they have been in use since the earliest days of computer design. The earliest microcomputer busses were designed to be compatible with a single type of processor and perform only rudimentary functions. As the industry matured, busses became increasingly complex with greater capability and higher speeds. Before proceeding with a detailed look at some currently popular busses, it is necessary to have an understanding of the basics of the function of a computer bus and some of the important factors to consider when looking toward a bus solution to a given problem.

It is always important to keep a bus in perspective. It is just one tool, of a variety of tools, available to solve a problem. It is also only one part of the final tool that will provide the complete problem solution. Although busses have increasingly been specified to be general purpose and to fill a vast array of generalized requirements, all of them have performance limitations that should be understood by engineers and engineering managers before a design decision is made. The final decision about which bus to use and how the system is to be designed around a given bus is a complex decision and is not algorithmic. The factors involved with such a decision are constantly changing and are not based on current requirements only (which may be poorly defined at the time of the decision) but also on cost, schedule, scalability, expandability, design margin, mechanical packaging, environmental considerations, and even political considerations within a project. The easiest (and fun) part of the job is usually after the basic and messy decisions are made. Invariably, the basic decisions are made on the basis of incomplete and inadequate information. Unfortunately, it is part of the job.

1.1 Bus Definition

Although, in the most general form, there are a vast array of different types and configurations of busses, for the purposes of this book a bus is

> A tool designed to interconnect functional blocks of a (micro) computer in a systematic manner. It provides for standardization in mechanical form, electrical specifications, and communication protocols between board-level devices.

Assuming that a design engineer has decided to use a microcomputer in a design, there are two general approaches that can be taken. It may be possible to achieve a satisfactory solution with a single-board computer, or a multiple-board solution may be indicated. When a multiple-board solution has been selected, consideration of the bus to be used should be started. Usually a bus is considered when the complexity or number of functions to be performed exceed the current capacity of state-of-the-art designs in single-board computers or the project design schedule does not permit a custom-board design. I am an advocate of using standardized busses whenever feasible in a design. Given the current flexibility and power of busses, there are only a few industrial or military problems that require the design of a custom or proprietary bus. Unfortunately, this decision can be political as well as technical; so technical judgment may not always prevail in such decisions.

1.2 Design Trade-Off Assessment Factors

There can be a variety of advantages to using a standard bus in any design. Design time can be reduced because of the following:

1. Few or no boards to custom design
2. Numerous vendors from which to select boards
3. "Custom" ICs available for bus interface
4. Board-level communications specifications done
5. Reduced troubleshooting time

1.2 Design Trade-Off Assessment Factors

Design and project costs can be reduced because of the following:

1. No bus specifications to develop
2. Reduced debugging and build time
3. Fewer engineering hours in hardware development

All it not perfect, however. There are disadvantages to using a bus in a design. Costs will be higher than a single-board computer implementation, although you get more power and performance. As a purchaser of board-level products, a design will be at the mercy of vendors over the product life. A given vendor may discontinue a part on little notice. (As all vendors say in fine print: "Specifications are subject to change without notice.") Availability problems may produce production slippages. All of these manufacturing problems are certainly not unique to board-level products but should be considered by any designer, as second sources on board-level products can be quite difficult to obtain. Even with these problems, the advantages generally outweigh the disadvantages and an increasing number of design engineers are using standard busses.

There are a few factors that should be considered by any engineer or engineering manager when determining whether to use a single-board computer or bus design. First on the list is complexity of the problem. Items used to estimate the complexity include

1. Estimate of the amount of memory required
2. I/O requirements (serial, parallel, special function)
3. File storage capability
4. Special processing functions
5. Parallel or pipelined processing requirements
6. Product support and development environment

Special processing functions may include preprocessors such as array processors, intelligent I/O, and high-speed specialized coprocessors.

Speed always seems to be high on the list of requirements. Any good systems engineer should know that the clock speed of the processor(s) in a system is not always the most important consideration when calculating system throughput; however, this number seems to always get the most press coverage. A more complete list of factors that determine system speed includes

1. Efficiency and implementation of algorithm
2. Multitasking overhead
3. Data transfer speed (bus speed)
4. I/O handling speed
5. Compiler and coding efficiency

One detail that must always be considered in a design is packaging. It can range from being a simple fact of life to an overriding concern in any given design. Factors that must be considered include form factor of the boards and bus, as well as environmental concerns. Most busses have a fixed form factor that must be accepted once the bus is selected. Some of the newer busses, such as VMEbus, support a variety of form factors, giving the packaging engineer some latitude in the design. Environmental factors may include

1. Temperature
2. Humidity
3. Radiation
4. Atmospheric dust and dirt
5. Altitude
6. Shock and vibration

Most commercial boards are designed for commercial or industrial environments, meaning a relatively dry humidity, controlled temperatures, minimal or nonexistent radiation, relatively low content of atmospheric dust and dirt, low altitudes, and low shock and vibration content. Invariably, a given requirement will be such that at least one item on this list will be a problem. All items on this list can be controlled with careful mechanical or packaging design. The packaging costs incurred by some design problems can far exceed the board, chassis, and component costs.

Some designs also have severe restrictions on power consumption. Most bus implementations are not extremely frugal on power consumption, but there has been an increasing awareness of the problems introduced by excessive power consumption in board-level designs. Power supplies become large, temperatures elevate causing poor reliability, and packages designed to remove large amounts of heat become expensive. The advent of high-speed CMOS devices has forced a reduction in the use of power-hungry NMOS, and board-level power consumption has dramatically dropped. Any design having poor cooling capability or that must be operated from batteries or that must be quite small should be examined for power consumption and possible thermal problems.

1.3 Signal Groups and Uses

Nearly all microprocessors and computer busses have four generic groups of signals. The details of implementation, usage, and capability of these groups of signals vary tremendously among realizations. A VMEbus design, for example, has much greater capability (and cost) than STD bus in general. This statement does not mean that STD bus has been superceded by VMEbus for all applications. But there is a range of performance and cost and the bus selected for a given design should be matched in performance and cost to the system requirements.

The four groups of common signals are

1. Data
2. Address
3. Control
4. Power

The data bus provides for the passage of data between external (board-level) devices and the bus master. In this book they will usually be labeled $D_0, D_1, D_2, \ldots D_n$ with the numeric suffix indicating the line number, that is, bit 11 is carried on D_{11}. The number of bits on the bus or processor determines the "size" of the bus or processor. For example, the Motorola 68020 has 32 data bus lines and is therefore a 32-bit processor. The data bus is also commonly used to carry interrupt vector information.

The address bus provides for the passage of address information between the bus master and external (board-level) devices. It is generally labeled $A_0, A_1, A_2, \ldots A_n$ with the numeric suffix indicating the address line number. The number of bits in the address bus (or the "width" of the address bus) determines the addressing range of the bus. For example, the Intel 8086 has a 20-bit-wide address bus, so the addressing range of the bus is 2^{20} bytes with the highest address being $2^{20} - 1$. This corresponds to the hexidecimal address of 0FFFFFFh, which is 1,048,576 bytes of memory or, more commonly, 1 Mbyte of memory. The address bus also normally carries port addresses for those processors supporting I/O instructions and may also be used as a part of the control bus for certain busses.

The control bus is the most variable part of any given bus type. It is also the group of lines that gives the most flexibility and power to a bus. Some of the functions carried out by the control bus may include power failure handling, interrupt processing, vector passing, multiple master arbitration, data transfer handshake, and similar functions. The number of lines that fit into this category may range from a few (in the case of the simpler STD-type bus) to several dozen for the more complex busses such as MULTIBUS II or VMEbus. Some functions may be supported minimally on some busses and extensively on other types. A systems designer or engineer must know and understand in detail the capabilities of each bus in order to successfully use its features and understand how it can help solve a design problem. Subsequent chapters contain considerable detail on the operation of the control features for each bus covered in this book.

The power bus is probably the simplest group of bus signals. But if insufficient attention is paid to the design of the backplane and power requirements for the system, it can be a source of problems. Most busses require a number of supply voltages at relatively high currents. Most busses support at least +5 VDC, +12 VDC, −12 VDC, and −5 VDC, with the −5 VDC rarely implemented in actual practice. Extensive use of power planes and multilayer backplane design are common in larger, more complex busses. The simpler busses may simply use two-layer routed power lines.

1.4 Transmission-Line Concepts

If you are a user of commercially produced bus systems, the information in this section is of background interest only. If you are a designer of a custom bus or boards using a standard bus, you should be aware of the basic tenets of transmission-line effects. Even for users, occasional failures will occur in systems that are the result of transmission line principles being neglected. As the speed with which systems operate becomes higher, the effects caused by long transmission lines, improper terminations, and mismatched distributed lumped constant loads become more critical. It is important to realize that all transmission-line problems have one common denominator, which is improperly controlled line and load impedances. Unfortunately, due to the manner in which busses operate and are configured, it is usually not feasible to adopt an ideal solution to the problem. Fortunately, due to the forgiving nature of most digital devices, it is only necessary to come reasonably close to an ideal solution.

In this section, digital signals are treated as analog signals and are modeled using analog techniques.

There are some symptoms that a system will exhibit that indicate a failure is being caused by transmission-line problems. A partial list of typical symptoms might include:

Symptom 1 When a program is executed in a target system, the system fails when accessing a certain board. Tests run independently on the board indicate that the board is operating correctly. Examining control, data, or address signals input to the board from the bus shows distortion with evidence of ringing and/or overshoot on rising and falling edges.

Symptom 2 A given board appears to fail during execution of a target system code. Making changes in the code so that the manner in which the board is accessed by the software "fixes" the problem but the code changes should have had no effect on the operation of the board. Changing the order in which the board is accessed or placing software delays at certain places in the code are examples of this.

Symptom 3 A system appears to be operating correctly until a modification is made to the system hardware configuration, such as adding a board or several boards. A previously working board suddenly begins to exhibit intermittent failures or stops working altogether.

Symptom 4 A board appears to be "slot sensitive" for no apparent reason. A given board works when installed in one slot but then stops working when installed in another slot.

Note that all of these symptoms could have a problem source other than transmission-line problems but these types of symptoms indicate that transmission-line problems should be on the list of potential causes.

A clad run or connection on a board becomes a transmission line when the length of the line becomes long enough to induce a delay time in the propagation of a digital signal that is a significant part of the rise time, fall time, or width of the signal being propagated. To make things a little (or a lot) clearer, all electrical signals propagate with finite speed. The speed is determined by the medium through which the signal is propagating. For a signal on a 10-mil-wide clad run on a glass epoxy (FR-4), 0.062-in.-thick printed circuit board with ground plane on the back side, propagation speed is about 2.0 ns/ft. Typical TTL high-current drivers have rise and fall times of about 1.0 ns. If a clad run is long enough to require more than about 0.50 ns for a signal to propagate, there is a potential for failure of the logic due to transmission-line effects. This corresponds to a distance of about 3.0 in. These numbers are somewhat arbitrary, and individual designs may be more or less tolerant than illustrated here, but the principle remains sound.

A transmission line has a given characteristic impedance. If the line impedance changes from one characteristic impedance to another, as a signal propagates down a line, part of the power is transmitted forward on the line and a reflected signal is generated traveling in the reverse direction on the line. The composite signal is the instantaneous sum of all of the incident and reflected signals on the line. The magnitude of the reflected signal is dependent upon the magnitude of the mismatch. Typically, in a digital system the point where a transmission line is mismatched most severely is at the load or receiver and in any particular system there may be several loads in the system producing multiple incident and reflected signals. It is also important to note that all incident and reflected signals are still subject to line losses, so a reflected signal observed at a transmitter will be smaller than the reflected signal observed at the point of reflection. All real transmission lines also attenuate high-frequency components of a signal more than lower frequency components, so sharp transitions tend to be "smoothed" or lost on both transmitted and reflected signals.

With these basic ideas in mind, printed circuit boards fall into two broad categories of design. Two-layer boards are inexpensive and easily designed but limit the designer to only two connection layers. They are therefore limited in the clad density such a design can support. Multilayer boards are more expensive to design and fabricate but vastly increase the clad density. The number of layers in a given design can easily exceed 12. (There are additional wiring techniques such as wire wrap, multiwire, and others; however, this discussion is limited to traditional types of printed wiring board fabrication techniques.) The additional layers also allow a designer to make extensive use of power and ground planes to reduce power supply noise-induced failures and line crosstalk and produce very closely controlled transmission-line impedances. Typical two-layer designs in a backplane use one layer for the signal and the second layer for a signal ground. Any power supply plane is a signal ground. Running a signal trace on one side of a board and ground on the second produces a layout cross section shown in Figure 1-1. This type of layout is called a microstrip transmission line. A multilayer board will typically have signal conductors interleaved with signal grounds, as shown in Figure 1-2. This type of layout is called a stripline transmission line.

A microstrip transmission line has the advantage of slightly faster propagation speeds than a stripline design and has a line characteristic impedance described by Equation (1.1):

$$Z_0 = \frac{87}{(E_r + 1.41)^{0.5}} \ln \frac{5.98 T_b}{0.8 W_c + T_c} \qquad (1.1)$$

where

E_r = Board material relative dielectric constant
T_b = Board thickness
T_c = Clad thickness
W_c = Clad run width

Note that this approximation assumes that electric field fringe effects are minimal. This is achieved by making the width of the signal ground plane at least four times the width of the clad and ensuring that the signal clad is centered above the signal ground plane. Most epoxy-based board materials have E_r values ranging from 3.5 to 5.5, while ceramics have higher dielectric constant values. If the dielectric constant is not known for a given board material in a microstrip configuration, it can easily be measured. By driving a clad run about one-half meter long with a fast-rise pulse

1.4 Transmission-Line Concepts

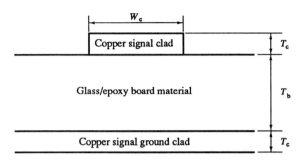

Figure 1-1 Microstrip transmission line.

generator and measuring the time difference between the rising edge of the incident pulse at the transmitter and the load, the E_r value can be calculated from Equation (1.2).

$$E_{cr} = [(300.0E + 6)L_c D]^{0.5} \tag{1.2}$$

where

E_{cr} = Composite relative dielectric constant
L_c = Length of clad run under test (meters)
D = Delay between transmitted and received signal

In a microstrip configuration, the relative dielectric constant is not simply the dielectric constant for the board material, as the clad run has board material on one side and air on the other so a composite or effective dielectric constant must be used. Equation (1.1) already compensated for the composite dielectric material effect and the board material dielectric constant should be used. Equation (1.2) calculates the composite dielectric constant, however. The thickness of 0.5 oz. copper clad board is

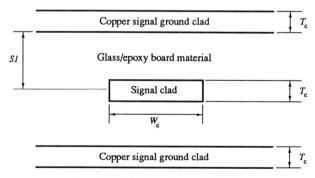

Figure 1-2 Stripline transmission line.

$T_c = 0.0014$ in. Most two-layer boards are 0.062 in. thick. Using an $E_r = 4.0$ and $W_c = 0.010$ in., the impedance of the transmission line is calculated from Equation (1.1) as

$$Z_0 = 137.5 \; \Omega$$

This is a typical characteristic impedance for many microstrip backplane lines. Some backplanes are designed and then subsequently conformally coated. Conformal coating increases the composite dielectric constant of the board. For many applications, this increase may not be significant enough to cause failures, but designers should be aware that the potential is present for failures to occur.

Microstrip conductors are completely embedded in the board material. The characteristic impedance of the line may be estimated from Equation (1.3):

$$Z_0 = \frac{60}{E_r^{0.5}} \ln \frac{8S_1}{0.67 \, (\pi) \, (0.8W_c + T_c)} \tag{1.3}$$

where

E_r = Relative dielectric constant of board
S_1 = Separation distance between board layers
π = 3.14159
W_c = Width of buried clad run
T_c = Thickness of buried clad run
Z_0 = Line characteristic impedance

The preceding equation is reasonably accurate if S_1 is at least three times T_c and W_c is less than half S_1. Equation (1.2) may be used to experimentally determine E_r if a board dielectric constant is not known. Typical propagation delays for FR-4 material are about 2.25 ns/ft. Typical stripline delays are about 1.75 ns/ft. Using $E_r = 5.0$, $T_c = 0.0014$, $W_c = 0.008$, $S_1 = 0.015$, and solving for Z_0 using Equation (1.3):

$$Z_0 = 53.4 \; \Omega$$

This also is a typical value for a stripline design. Note that in practice many (although not all) stripline designs tend to produce lower characteristic impedances.

1.4.1 Transmission-Line Terminations and Reflections

The basic transmission-line configuration shown in Figure 1-3 shows a driving source (which in this case produces a step voltage driving function), transmission line of characteristic impedance Z_0, and a termi-

1.4 Transmission-Line Concepts

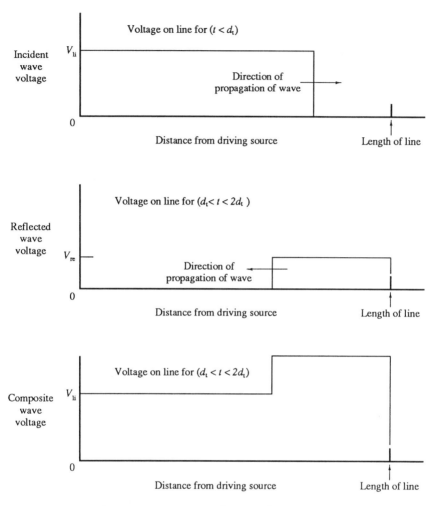

Figure 1-3 Transmission-line waveform ($R_l > Z_0$).

nating or load resistor R_l. The current produced in the line is V_s/I_1, where V_s is the magnitude of the voltage step and I_1 is the transmission-line current. The step in line voltage occurs at time t_0, with the step occurring at the load resistor at time $t_0 + d_t$, where d_t is the group delay of the transmission line. When the step voltage and current waveform occur at the load resistor, a current of I_1 is generated in the load resistor if the value of the load resistor is the same as the transmission line characteristic impedance Z_0. Under these conditions, there is no reflected current or voltage waveform on the line and the line is said to be matched. This is the ideal condition under which backplanes and board transmission lines operate, as this produces the "cleanest" waveforms. The only effect the transmission line has is to delay the arrival of the step by the group delay of the line d_t. When a load impedance is not the same or "matched" to the line impedance, the initial line current is

$$I_l(t_0) = V_s/Z_0$$

But the final steady state current on the line is determined by the load resistance (assuming the ohmic resistance of the line is small) and is

$$I_l(t_0++) = V_s/R_l$$

where t_0++ is the time required for the line to reach the steady state value. Because the transmission line produces a different current on the line and the load requires a different current, a reflected wave is produced traveling in the opposite direction when the incident wave arrives at the load resistor. Since Kirchhoff's current and voltage laws must be satisfied at every instant in time,

$$I_{li}(t_0 + d_t) + I_{re}(t_0 + d_t) = I_{ld}(t_0 + d_t) \tag{1.4}$$

and

$$V_{li}(t_0 + d_t) + V_{re}(t_0 + d_t) = V_{ld}(t_0 + d_t) \tag{1.5}$$

where

I_{li} = Current in transmission line
I_{re} = Current reflected on transmission line
I_{ld} = Current in load resistor
V_{li} = Voltage of step generator on line
V_{re} = Voltage reflected on transmission line
V_{ld} = Voltage across load resistor
t_0 = Time generator produced a step of V_s
d_t = Delay of transmission line

For brevity, the time factor will be eliminated in future equations but it is assumed that the point at which the incident wave arrives at the load is used in the following equations. A simple application of Ohm's law produces

$$I_{ld} = V_{ld}/R_l$$

Substituting Equation (1.5) for V_{ld} produces

$$I_{ld} = (V_{li} + V_{re})/R_l \tag{1.6}$$

1.4 Transmission-Line Concepts

and since
$$I_{li} = V_s/Z_0 \tag{1.7}$$
and
$$I_{re} = -V_{re}/Z_0 \tag{1.8}$$
solving the equations for V_{re},
$$V_{re} = V_{li}(R_l - Z_0)/(R_l + Z_0) \tag{1.9}$$

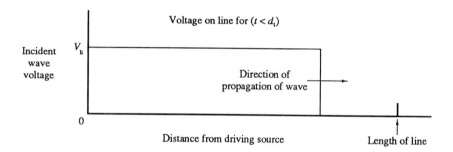

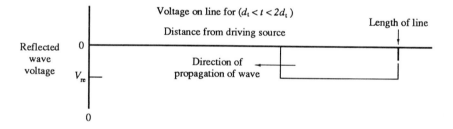

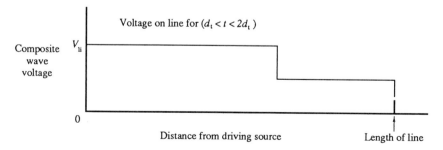

Figure 1-4 Transmission-line waveform ($R_l < Z_0$).

The term $(R_1 - Z_0)/(R_1 + Z_0)$ is called the reflection coefficient and represents the magnitude of the reflected wave as a function of the incident wave, load resistance, and characteristic impedance of the line. The sequence of events for the voltage and currents traveling on a transmission line having a terminating resistance greater than the characteristic impedance is shown in Figure 1-3. The voltage and currents on a transmission line having a terminating impedance less than the characteristic impedance are shown in Figure 1-4.

If the source impedance is not matched to the line characteristic impedance, the reflected wave from a mismatched load resistance acts as an incident wave. This in turn produces a reflected wave from the driving point. That reflected wave travels down the line toward the load and is again reflected by the load. At each reflection, the reflected wave is altered by the reflection coefficient and is attenuated by the losses on the line during its passage on the line. This multiple reflection process produces

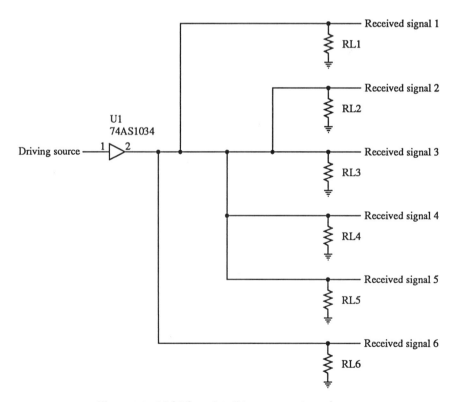

Figure 1-5 Multiple or "star"-type connection scheme.

1.4 Transmission-Line Concepts

the characteristic "ringing" effect to a step input for which mismatched transmission lines are noted.

It is rare that lines are terminated in a purely resistive load but are normally terminated in a parallel combination of resistance and capacitance with the capacitance being an inherent (undesirable) property of the receiver input or multiple lines fed in a starlike configuration from one driving source with multiple terminations (as shown in Figure 1-5). Capacitive loads increase the rise and fall times of the load signal and thus increase the overall delay from when a signal is sent by a transmitter and detected by a receiver in addition to the simple d_t delay of the transmission line. When there are multiple terminations configured in a star, each

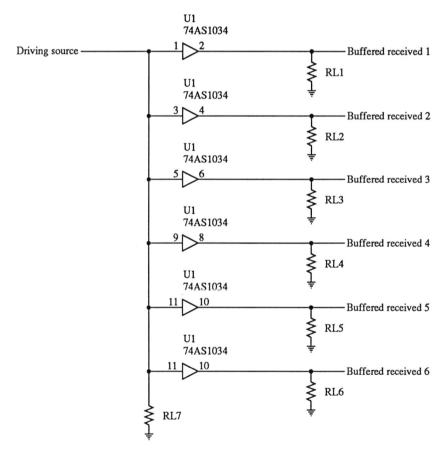

Figure 1-6 Buffered driver connection scheme.

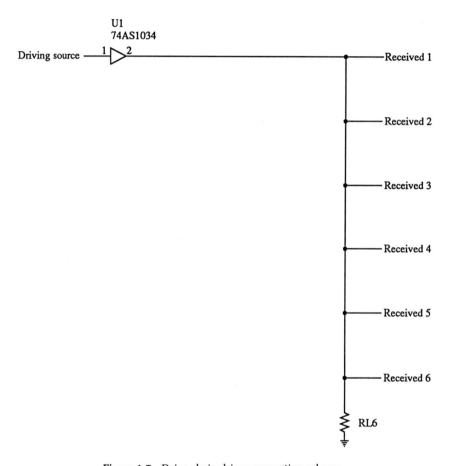

Figure 1-7 Daisy-chain driver connection scheme.

line and termination produces its own characteristic reflection wave which algebraically sum at each instant in time at any particular point on the transmission-line system. The effects of each one may be considered independently, with the resultant being the sum of the individual incident and reflected components. As the number of star fed lines and termination taps on a connection system increases, the analysis problem increases quite quickly. Long transmission lines perform best when isolated by buffers or when fed multiple loads in a series or daisy-chain manner, as shown in Figures 1-6 and 1-7, respectively.

2

A Comparison of Busses

2.0 Limitations

Performing a relative comparison between busses is a dangerous business at best. Any comparison, like any general statement, must be viewed with the understanding that it is the opinion of one person. There are also limitations in the techniques employed by which the general statements about performance are derived. Certain features of a bus may be overlooked, while others may figure prominently.

Any given application requires performance goals and other constraints in terms of environmental conditions and physical packaging that may drive a designer toward a given physical embodiment even though the raw computational and architectural power may not be present to the degree desired. Hardware architecture can be reshaped around other system constraints, and software can be implemented in a myriad of ways such that a given set of requirements for an application can be filled.

There is no one solution to any problem that I have ever encountered. There are a number of techniques that will work; however, not all solutions work as well as other solutions. Some solutions are "elegant." They require few parts, minimize the software requirements, are low in cost, and are implemented in a direct manner. A direct manner means that the design is simple, easy to follow, and easy to debug and change to new requirements. Some solutions are "messy." The hardware architecture is redundant and convoluted, resulting in high costs. The software required to operate a badly designed system also becomes convoluted and complex. With this, the time required to debug and change any part of the design escalates.

The ability to see the elegant as opposed to the messy solution comes with experience, aptitude, and knowledge. Because of the complexity of any design process, there are no widely accepted algorithmic techniques by which designs are implemented. There are a variety of tools and techniques that tend to push a designer in the direction of a well-structured design and the process is tending to become more procedural. However, this process still fits into the category of being creative.

After all of these words, the bottom line is that this chapter is a tool. It is designed to be used by the designer who is unfamiliar with the busses discussed in this book as a first pass on the selection of a bus or several busses that may be used for a given application. For all its problems and limitations, it still has value. (I think.)

2.1 A Basis for Comparison

There are a few basic functions that are performed by busses. Not all busses perform all of these functions and some implement the functions more powerfully than others. The basic functions used for a bus comparison in this chapter are

1. *Data transfers*   One of the most fundamental requirements of any bus is the ability to move data from slaves to masters, masters to masters, and masters to slaves. There are several techniques that are supported by different busses, including block transfers, byte, word, and long-word transfers.

2. *Interrupts*   The process by which a master is directed to "immediately" stop performing a given task and begin a (usually limited) new task is supported by virtually every microprocessor manufacturer in existence. It is a very commonly supported feature in nearly every bus. Each bus has different capabilities and techniques for handling interrupts.

3. *Multiple masters*   Most new busses have the ability to support multiple masters. There are actually two elements that must be examined in busses that support multiple masters. There is an arbitration phase in which the identity of the new master is determined among several masters that may be requesting the bus at the same time and the bus exchange process by which the bus control or ownership is moved from the old to the new master.

4. *Utility functions*   This is difficult to quantify and is the most variable part of any bus specification. It includes such features as error handling, power failure detection, and clocks.

In addition to the preceding performance functions, each bus will be evaluated for board real estate or size.

Any consideration as to power requirements is missing from this chapter. Nearly all busses use +5, +12, −12, and, optionally, −5 VDC. This is a detail that is important but not a consideration from a more top-level view of bus capability. I have included the ability to support battery backup as a part of the utility functions.

2.2 Which Busses

There are a large number of busses available to a designer. However, in the interests of completing this text within my lifetime, I have limited this discussion to a few commercial and industrial busses that are currently in common usage. They are the busses discussed in detail in the following chapters of this book. There may be other busses for a given application that are better suited than those discussed here, and the absence of a bus from this book should not be interpreted as an indication that the bus is inadequate in any way. The busses are

Common bus name	IEEE specification
MULTIBUS I	IEEE-796
MULTIBUS II	IEEE-1296
VMEbus	ANSI/IEEE-1014
NuBus	IEEE-1196
PC/XT/AT bus	None (EISA)
STD bus	IEEE-P961/D1 (proposed)

2.3 Bus Comparison

The following tables summarize the performance of each of the busses in key performance areas using the categories outlined above.

2.3.1 Data Transfers

A tabular comparison of data transfer capability for each of the busses is shown in Figure 2-1. Key elements in the comparison are the transfer type (synchronous/asynchronous), maximum transfer rate in Mbytes/s, the number of data lines, width of address bus, and types of transfers supported.

	Transfer type (Sync/Async)	Transfer rate (Mbyte/sec)	Number data lines	Multiplexed address data	Width address bus	Transfer types supported				Notes
						Byte	Word	Long word	Block	
Multibus I	Async	10.0	16	No	30	X	X			
Multibus II	Sync	40.0	32	Yes	32	X	X	X	X	24-bit transfers supported
VME bus	Async	40.0	32	No	16/24/32	X	X	X	X	Dynamic addressing range
NuBus	Sync	40.0	32	Yes	32	X	X	X	X	
PC/XT/AT bus	Sync	Processor dependent	PC/XT-8 AT-16	No	PC/XT-20 AT-24	X	AT			
STD bus	Sync	Processor dependent	8	Yes	16/20/24	X				

Figure 2-1 Comparison of bus data transfer capability.

	Number lines	Directed interrupts	Interrupt types		Notes
			Bus vectored	Nonbus vectored	
MULTIBUS I	8	No	Yes	Yes	
MULTIBUS II	N/A	Yes	No	No	Interrupts use message handling facilities
VMEbus	8	Yes	Yes	No	
NuBus	0	No	Yes	No	Interrupts handled using event transactions
PC/XT/AT bus	PC/XT-5 AT-12	No	Yes	No	
STD bus	1	No	Yes	No	

Figure 2-2 Comparison of bus interrupt handling capability.

	Multiple masters	Arbitration algorithms					Notes
		Single level	Multiple level	Rotating	User defined	Fair	
MULTIBUS I	Yes	X	X				
MULTIBUS II	Yes		X			X	Separate arbitration bus pipelined operation
VMEbus	Yes	X	X	X	X		
NuBus	Yes					X	
PC/XT/AT bus	PC/XT-No AT-Yes						DMA supported
STD bus	No						DMA supported

Figure 2-3 Comparison of bus master capability.

2.3 Bus Comparison

2.3.2 Interrupts

A comparison of interrupts and interrupt handling capability for each of the busses is shown in Figure 2-2. Key elements in the comparison of interrupts include number of interrupt lines, ability to direct interrupts, and types of interrupts supported (bus vectored and nonbus vectored).

2.3.3 Multiple Masters

A summary of multiple master handling capability is shown for each of the busses in Figure 2-3. Key elements in this comparison are the ability to support multiple masters, the arbitration scheme(s) supported, the arbitration time, and arbitration algorithms.

2.3.4 Utility Functions

There are a number of utility functions performed by each bus; however, the comparison is performed strictly on the ability of the bus to support the following:

1. Power failures
2. Bus errors
3. Battery backup capability

	Power failure handling	Bus error detection	Battery backup capability	Notes
Multibus I	Yes	No	No	
Multibus II	Yes	Yes	Yes	Parity and transfer error detection
VME bus	Yes	Yes	No	
NuBus	Yes	Yes	No	
PC/XT/AT bus	No	(See note)	No	Bus error detection limited to parity
STD bus	Yes	No	Yes	

Figure 2-4 Comparison of bus utility functions.

	User defined backplane pins	Board dimensions (in.)			Usable board area (in.²)			Number connectors			Number pins/ connector	Notes
		Config. 1	Config. 2	Config. 3	Config. 1	Config. 2	Config. 3	Config. 1	Config. 2	Config. 3		
Multibus I	No	6.7 x 12.0			80.4			1			86	Some utility on P2 connector
Multibus II	Yes	8.7 x 9.2	8.7 x 14.4		77.8	124.0		2	3		96	Other configuration supported
VME bus	Yes	3.9 x 6.3	6.3 x 9.2		24.8	58.0		1	2	3	96	Other configuration supported
NuBus	Yes	4.0 x 12.86	11.0 x 14.4		50.8	158.8		1	3		96	User of user defined pins controlled by spec
PC/XT/AT bus	No	4.2 x 13.15	4.2 x 6.0	4.5 x 13.15-AT	50.5	22.6	54.4	PC/XT-1	AT-2		J1/J8-62 J10/J16-36	
STD bus	No	4.5 x 6.5			27.9			1			56	Ejector reduces available space

Figure 2-5 Comparison of bus mechanical features.

2.3 Bus Comparison

Individual busses support a number of additional functions on their utility lines. However, they are often used in conjunction with other operations being performed on the bus.

A summary of each of the utility functions above supported by each bus is shown in Figure 2-4.

2.3.5 Mechanical

Board size and the maximum number of boards that can be fit into a given card rack are often a consideration in a design. A summary of board size and card count is given in Figure 2-5.

3

Software/Hardware Interactions

3.0 What and Why, or "A Touch of Philosophy"

Having worked in this industry for longer than I care to admit, I have interviewed a number of engineers and technicians for jobs and been in the unfortunate position of having had to make recommendations to my boss regarding their capabilities. We seem to have entered into the era of the "specialist." There are the software engineers and the hardware engineers (and analog engineers and control systems engineers and systems specialists and . . .). This artificial division, encouraged by a number of communities, has a lot of benefit, but it has a drawback in that, in a day and age of shrinking budgets and high labor costs, it is very difficult to find the magic individual who is capable of bridging the gap between hardware and software. So, this chapter is a very brief effort to make that connection and make a specialist into a little bit of a generalist. In any case, I have always found it useful to write and debug my own test and debug software for the boards I have designed. If you're not interested in this kind of stuff, or feel comfortable with your understanding, skip the chapter. The material presented here gives a very top-level view of the interaction and how software produces a given set of bus signals. There are specific examples, so it may be useful to skim this chapter until a more detailed understanding of the busses used in the examples is obtained and then spend more time on the examples.

This chapter is limited to dealing with the movement of data or instructions on the bus. Other bus operations such as interrupt handling, error processing, power failure, and others are not dealt with here in the interest of simplicity.

3.1 The Big Picture

From the perspective of bus operations, a bus master will perform bus operations based on the execution of instructions. In general, the instructions require the fetching of other instructions and the reading or writing of data to memory locations. Some busses also support I/O (input/output) instructions. In the process of executing instructions, a bus master may also be required to perform some operation on some data. The data may be read from memory and transferred to the master where some operation is performed on the data or data may be written by the master and transferred to memory or an I/O device called a slave. A bus master is a device that initiates all bus operations. A bus slave is a device that responds to bus operations initiated by the master.

All instructions consist of binary values that are loaded into the central processor's instruction interpretation unit from memory. Each processor has a "primitive" set of instructions that it is capable of executing and each instruction set is unique to the processor. No two types of processors execute the same primitive instructions but there is a great deal of similarity between the types of instruction each processor can execute. All processors can perform arithmetic operations, data movement operations, string operations, binary (or Boolean) operations, and branch operations with varying degrees of complexity and sophistication.

3.1.1 Assemblers

As assembler is a program that performs a reduction operation on a text file, producing the executable binary file needed to produce the native instructions for a given processor. So the input to an assembler program is a text listing of instructions specific to the processor. Once a program or software module has been assembled, the output is a binary file capable of being loaded into memory and executed by the processor.

Two examples of an exerpt from an assembly language source module are shown in Figure 3-1. This was written for a MULTIBUS board and executed on an 8086 family machine and intended to control a board containing a real-time clock (RTC). This module is a collection of subroutines that are callable from a Pascal program. A subroutine is a set of instructions that can be called repeatedly by a master program or other subroutine. A subroutine implements a function that may be commonly used. Pascal is an example of a high-level language.

The GetTime procedure contains a number of assembly language instructions. In brief, some of the more important of these are

3.1 The Big Picture

```
; ******************
; Function GetTime (Var Hours, Var Minutes, Var Seconds: Integer) : Boolean;
;    Returns the current hours, minutes and seconds time.  Function returns
;    true if no problems detected or false if problem found.
; ******************
GETTIME         PROC    NEAR
;
; GET ADDRESSES OF VARIABLES
;
        POP     RETADR                  ;SAVE RETURN ADDRESS
        POP     OFFSEC                  ;GET OFFSET OF SECONDS
        POP     ES
;
; CHECK STATUS TO DETERMINE IF RTC CAN BE READ
;
;       PUSH    DS                      ;PREPARE TO CALL TO RTCSTATUS
;       MOV     AX, OFFSET VRT
;       PUSH    AX
;       PUSH    DS
;       MOV     AX, OFFSET UIP
;       PUSH    AX
;       CALL    RTCSTATUS               ;CHECK STATUS OF RTC
;       CMP     AX, FALSE
;       JE      OUT1
;
; GET TIME SINCE NO ERROR CONDITIONS
;
        MOV     DX, REGSEC              ;GET SECONDS DATA
        MOV     AX, 0
        IN      AL, DX
        MOV     BX, OFFSEC              ;PLACE DATA IN VARIABLE
        MOV     ES: BYTE PTR [BX], AL
;
        POP     OFFMIN                  ;GET OFFSET OF MINUTES
        POP     ES
        MOV     DX, REGMIN              ;GET MINUTES DATA
        MOV     AX, 0
        IN      AL, DX
        MOV     BX, OFFMIN              ;PLACE DATA IN VARIABLE
        MOV     ES: BYTE PTR [BX], AL
;
        POP     OFFHR                   ;GET OFFSET OF HOURS
        POP     ES
        MOV     DX, REGHR               ;GET HOURS DATA
        MOV     AX, 0
        IN      AL, DX
        MOV     BX, OFFHR               ;PLACE DATA IN VARIABLE
        MOV     ES: BYTE PTR [BX], AL
;
        MOV     AX, TRUE                ;SET FUNCTION RETURN OK - NO ERROR
OUT1:
        PUSH    RETADR                  ;RESTORE RETURN ADDRESS
        RET

GETTIME         ENDP
```

Figure 3-1 Assembly language example: RTC MULTIBUS GetTime Pascal procedure.

POP	Pulls a value off the stack and stores
PUSH	Places a value onto the stack
MOV	Moves a data value from a register to memory or memory to a register
CALL	Executes another procedure and returns
JE	A conditional branch instruction (branch if equal)
IN	An I/O instruction that moves data from an I/O device to a register

Note that any line beginning with a ";" indicates that the line is either a comment or it is "commented out." This means the line has no significance to the assembler.

The GetTime procedure contains four sections. The addresses of the variables are stored on the stack when the procedure is called and these addresses must be retrieved. The first three POP instructions perform this function. Note that the function must know where to execute the next instruction in memory after this procedure is complete. Pascal places the return address of the next instruction to be executed after completion as the last value on the stack. This is the first value "POPed" from the stack upon execution of the GetTime (or any other procedure). The next section (commented out) checks the status of the RTC for error conditions. Finding no errors, the third section reads the seconds, minutes, and hours from the RTC. The values read are placed in the addresses of the variables declared by the Pascal calling procedure whose addresses are stored on the stack. The last section restores the return address to the stack and returns (RET) to the calling procedure.

The GetDate procedure works in a manner similar to the GetTime procedure. (See Figure 3-2.) The registers read a different address but the procedure is otherwise very similar.

The specific addresses of the RTC registers are declared at the beginning of the assembly language module.

3.1.2 High-Level Languages

Because assemblers are specific to a processor or machine, high-level languages have been developed to make programs or "code" portable. Portable code is capable of being moved to nearly any machine, and can be compiled and executed on that machine with no changes to the code. High-level languages are text source files that form the input to a compiler. The compiler takes the text file and generates either an assembler text output file or executable binary file that may be directly loaded and executed by the machine. If an assembler file is generated, it must be

```
; ******************
; Function GetDate (Var Year, Var Month, Var WeekDay, Var MonthDay: Integer)
;                  : Boolean;
;   Returns the year, month, weekday, monthday.  Returns function true if
;   no RTCStatus returns true.
; ******************

GETDATE         PROC    NEAR
;
; GET ADDRESSES
;
        POP     RETADR                  ;RETURN ADDRESS
;
; CHECK STATUS TO DETERMINE IF RTC CAN BE READ
;
        PUSH    DS
        MOV     AX, OFFSET VRT          ;PREPARE TO CALL RTCSTATUS
        PUSH    AX
        PUSH    DS
        MOV     AX, OFFSET UIP
        PUSH    AX
        CALL    RTCSTATUS               ;CHECK STATUS OF RTC
        CMP     AX,FALSE
        JE      OUT2
;
; GET DATE SINCE NO ERROR CONDITIONS
;
        POP     OFFMDAY
        POP     ES
        MOV     DX,REGMDAY              ;GET MONTH DAY DATA
        MOV     AX,0
        IN      AL,DX
        MOV     BX,OFFMDAY
        MOV     ES: BYTE PTR [BX], AL   ;PLACE DATA IN VARIABLE
;
        POP     OFFWDAY
        POP     ES
        MOV     DX,REGWDAY              ;GET WEEK DAY DATA
        MOV     AX,0
        IN      AL,DX
        MOV     BX,OFFWDAY
        MOV     ES: BYTE PTR [BX], AL   ;PLACE DATA IN VARIABLE
;
        POP     OFFMO
        POP     ES
        MOV     DX,REGMO                ;GET MONTH DATA
        MOV     AX,0
        IN      AL,DX
        MOV     BX,OFFMO
        MOV     ES: BYTE PTR [BX], AL   ;PLACE DATA IN VARIABLE
;
        POP     OFFYR
        POP     ES
        MOV     DX,REGYR                ;GET YEAR DATA
        MOV     AX,0
        IN      AL,DX
        MOV     BX,OFFYR
        MOV     ES: BYTE PTR [BX], AL   ;PLACE DATA IN VARIABLE
;
        MOV     AX,TRUE                 ;SET FUNCTION RETURN OK - NO ERROR
OUT2:
        PUSH    RETADR                  ;RESTORE RETURN ADDRESS
        RET

GETDATE         ENDP

; ******************
```

Figure 3-2 Assembly language example—RTC MULTIBUS GetDate Pascal procedure.

subsequently assembled by a native machine assembler. All compilers are specific to the machine for which they are targeted; however, the source code used as the input to the compiler is portable.

There are a number of high-level languages currently in use. Some of the more popular general-purpose or scientific languages include "C," Pascal, ADA, and Basic. Each language is capable of performing the same basic operations as an assembler but with greater power and ease. All compilers and assemblers must be capable of supporting arithmetic operations, string operations, boolean operations, data movement, and branch operations.

An example of a Pascal source listing exerpt from a program used to test the assembly language module shown in Figure 3-1 is shown in Figure 3-3. Figure 3-3 is a "procedure." Pascal is a procedural language. Pascal programs consist of collections of procedures and functions that are culled from other procedures and functions that perform one basic task. Data may or may not be passed into the procedure and the procedure may or may not produce data output.

As can be seen in the listing of Figure 3-3, it is not necessary to be very familiar with Pascal to have a general understanding of the procedure's function. The procedure DispTimeDate gets the time and date from the RTC by calling two assembly language procedures called GetDate and GetTime. GetDate returns the current year, month, week day, and day of the month. GetTime returns the current hour, minutes, and seconds. The parameters returned by GetDate and GetTime are formatted and printed out in a usable format after checking for an error in getting the data.

3.2 Bus Interaction

Taking a closer look at Figure 3-1, after the POP instructions are executed, the first instruction to be executed is

MOV   DX,REGSEC

The variable REGSEC contains the address in memory of a variable called "Seconds." That address was passed to the GetTime procedure by the Pascal calling procedure and was placed on the stack. The stack is an area in memory used for the temporary storage of variables and addresses or "pointers." DX is a 16-bit register internal to the microprocessor. This instruction requires two bus operations. The first operation is to load the instruction into the instruction cue on the processor, and the execution of the instruction requires the movement of the address REGSEC into the

3.2 Bus Interaction

```
(* ********************************************************* *)
Procedure DispTimeDate;
(* Gets the time and date from the RTC and displays *)

Var
   Seconds         :       SecRange;
   Minutes         :       MinRange;
   Hours           :       HrRange;
   WkDay           :       WkDayRange;
   MoDay           :       MoDayRange;
   Month           :       MoRange;
   Year            :       YrRange;

Begin
   Writeln;
   Writeln;
   If GetDate (Year, Month, WkDay, MoDay) then
     Begin
       Writeln ('No error detected during date read..');
       Write ('Date read: ');
       Case WkDay of
         1: Write ('Sunday ');
         2: Write ('Monday ');
         3: Write ('Tuesday ');
         4: Write ('Wednesday ');
         5: Write ('Thursday ');
         6: Write ('Friday ');
         7: Write ('Saturday ');
       End; (* Case *)
       Writeln (Month:2,'/',MoDay:2,'/',Year:2);
     End
   Else
       Writeln ('Error detected during date read..');
   If GetTime (Hours, Minutes, Seconds) then
     Begin
       Writeln ('No error detected during time read..');
       Writeln ('Time read: ',Hours:2,':',Minutes:2,':',Seconds:2);
     End
   Else
       Writeln ('Error detected during time read');
   Write ('Press any key to continue..');
   Read (Dummy);
End; (* DispTimeDate *)
```

Figure 3-3 Pascal source listing DispTimeDate procedure—calls assembly language modules shown in Figures 3-1 and 3-2.

microprocessor's DX register. This operation might occur as shown in Figure 3-4 assuming the instruction is executed on a MULTIBUS system. It may be convenient to read Chapter 4 for a better understanding of the bus operation and signals.

The next instruction is

$$\text{MOV} \qquad \text{AX,0}$$

This instruction requires only one bus operation as the instruction loads the 16-bit hexidecimal value of zero into the microprocessor's AX register.

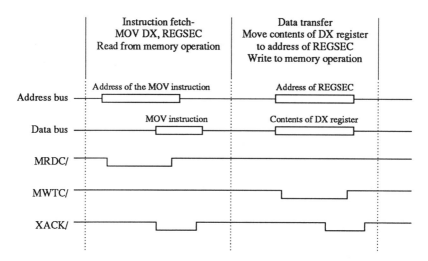

Figure 3-4 MOV DX, REGSEC timing diagram on MULTIBUS.

Since there is no data movement in or out of memory or I/O space, a single instruction fetch cycle is performed.

The operation of the instruction,

$$\text{IN} \quad \text{AL,DX}$$

requires two bus operations. The first, as usual, is the fetch of the instruction from memory and the second is an I/O instruction which goes to the I/O address pointed to by the DX register and moves the contents of that address into the low eight bits of the microprocessor's AX register, which is called the AL register. The execution of this instruction causes signals to appear on a MULTIBUS I system as shown in Figure 3-5. Note that VMEbus does not intrinsically support I/O instructions, so special provisions must be made in the design of the bus master that supports I/O instructions. Each such implementation is processor-specific.

By following the instructions step by step it is possible to determine every bus operation that must be performed to execute every assembly language instruction. When using high-level languages, there is usually more overhead associated with the operation of the language. There is also usually more overhead when the resultant code is "disassembled" or an assembly language source listing for a high-level language module is examined. The advantages of a high-level language are portability and ease of development. The disadvantage is usually efficiency of execution

3.2 Bus Interaction

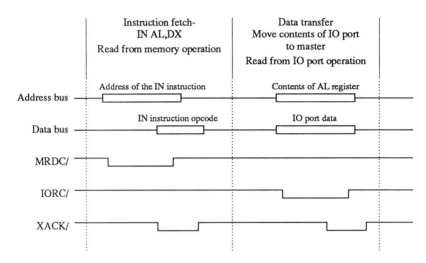

Figure 3-5 IN AL, DX timing diagram on MULTIBUS I.

in that there are often instructions in a general-purpose procedure that are not required for a given specific implementation. It is, however, possible to step through a high-level language module, examining each step to determine which bus operations must be performed in a manner similar to that shown here for the assembly language module.

4

MULTIBUS I

4.0 MULTIBUS I Overview

MULTIBUS I is fast becoming one of the "grandfather" busses in the microcomputer world. However, in spite of the age of the bus, it has a great deal of capability and is one of the most popular busses in existence. Since its inception in the early 1970s, other more powerful busses have been developed. However, MULTIBUS I was the first successful commercial bus to be designed that extended the bus's capability beyond the limited capability of a single microprocessor. It also was easily adapted to a variety of different microprocessors. At the time of its release, MULTIBUS I was one of the most powerful and capable commercial busses on the market and represented a major advancement in tools available to a design engineer. As the bus market has matured, more powerful boards have been developed for MULTIBUS I and the cost of the boards developed over the past few years has dropped, making the bus viable for a mid-performance product at a moderate price tag.

The original MULTIBUS standard developed by Intel actually supported four busses. Those busses are

1. MULTIBUS system bus
2. I/O expansion bus (iSBX)
3. Execution bus (iLBX)
4. Multichannel I/O bus

The mechanical outline of a MULTIBUS I board is shown in Figure 4-1. A photograph of a typical MULTIBUS board is shown in Figure 12-2a (p. 219). A system layout for a complex MULTIBUS system using all of the extension busses might be as shown in Figure 4-2. Note that there are two connectors used by the MULTIBUS board. The MULTIBUS system bus

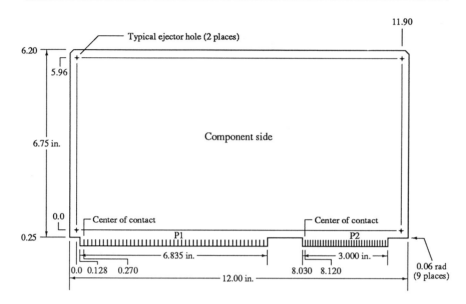

Figure 4-1 MULTIBUS I board outline and dimensions.

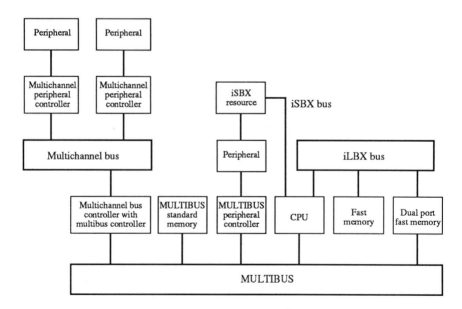

Figure 4-2 MULTIBUS I typical system.

has all signals implemented on the P1 connector or the large 86-pin card edge connector. The iLBX bus is implemented largely on the P2 connector. The MULTIBUS system bus will be covered in some detail in this chapter. The other busses will be mentioned only briefly. The MULTIBUS system bus supports a number of innovative features that were rare in microcomputers prior to the release of this bus, including

1. Ability to operate with multiple masters
2. Support of 8-bit and 16-bit devices simultaneously
3. Transfer rates up to 10 Mbytes/s
4. Twenty address lines (1 Mbyte address space)
5. Sixteen bidirectional data lines
6. Eight multilevel interrupt lines

After the bus had been on the market for several years, the Institute of Electrical and Electronics Engineers (IEEE) adopted the original Intel MULTIBUS specification with minor modifications as an IEEE specification called IEEE-STD 796.

4.1 iSBX Bus

The iSBX bus provides for on-board expansion of resources. The function of the iSBX bus is to provide a designer with the ability to tailor required functions and system performance with relatively low-cost module boards that can be plugged into single-board computers (SBCs) or expansion boards. The SBCs or expansion boards then have increased capability without using any of the bandwidth capabilities of MULTIBUS. The iSBX bus supports all of the traditional classes of lines of a computer bus, including address, data, interrupt, control, and power. Operations supported by the iSBX bus include I/O read, I/O write, direct memory access (DMA), and interrupts. The base or host board decodes the addresses and generates the selects for the iSBX module that is plugged into the host. The host must have iSBX interface capability designed into the board. There are two mechanical standards for iSBX modules. One is a single, wide board, 3.7 in. long, with a double wide standard also supported that is 7.5 in. in length.

4.2 iLBX Bus

The local bus extension (iLBX) is a special bus with a predefined special electrical protocol and mechanical interface specification providing for local memory expansion off-board without loss of execution speed. A single-board computer communicates to an external memory board over the iLBX lines. There may be up to five memory expansion boards in the expanded memory of a CPU board. The LBX memory is treated by the CPU as local memory. The principal vehicle for decreasing memory access times is achieved by the lack of bus arbitration that must be accomplished by a general-purpose bus interface. Up to 16 Mbytes of memory on five boards may be accessed by a single iLBX bus. There may be several iLBX busses in a single system. Additionally, some iLBX memory boards have dual port memory on the iLBX board, allowing the memory module to be accessed by the iLBX bus or by the system MULTIBUS. LBX module boards have the same form factor as MULTIBUS and communicate to the LBX master over the 60-pin P2 connector.

4.3 Multichannel I/O Bus

The multichannel I/O bus provides yet another means of expanding the I/O handling capability of a MULTIBUS system. This bus is specifically oriented toward decreasing the impact of burst-oriented devices on multibus system performance. Data transfers occurring from a high-speed burst transfer device can easily use all of the bandwidth capacity of a general-purpose bus such as MULTIBUS. Several burst devices such as hard disk drives can significantly reduce performance of a bus. When fully implemented, the multichannel bus uses dual port buffer memory. One port of the memory is accessed by the burst device by using the multichannel I/O bus, while the second port of the dual port memory is accessed by the MULTIBUS system bus. The multichannel I/O bus supports up to 16 Mbytes of addressable memory and can support up to 16 devices at distances of up to 15 m from the host machine. The bus will support data transfer rates of up to 8 Mbytes/s and will support both 8- and 16-bit data transfers. The Multichannel bus can also be used to link two MULTIBUS base system crates together. Multichannel transfers are asynchronous using a positive handshake protocol. Additionally, parity verification is used to ensure data integrity. Typical types of devices that may be linked via the multichannel bus include

1. Video frame grabber and frame store
2. Graphics terminals
3. Workstations
4. Array processors
5. Other MULTIBUS crates

4.4 MULTIBUS System Bus

The MULTIBUS I system bus supports a number of powerful features. Included in those features are

- Memory and I/O data transfers
- Direct memory access
- Generation and handling of multiple interrupts
- Direct memory addressing of up to 16 Mbytes
- Direct I/O addressing of up to 64 kbytes
- Support of 8- or 16-bit data transfers
- Data transfer speeds of up to 5E+6 transfers/s
- Support of multiple masters

All transfers on MULTIBUS are built on the master/slave concept of bus transfers. A MULTIBUS master is a device or board that initiates bus transfer requests and a slave is a device that responds to such requests. Traditional thought and implementation are such that a master is implemented on one board and slaves may each occupy a single board slot. With the increasing sophistication of system, it is more common to find both masters and slaves implemented on a single board or to find multiple masters on a single board.

In any case, the master/slave relationship is a cornerstone for the operation of MULTIBUS and most other types of busses. There is a handshake or transfer protocol for each type of data or control transfer that can occur on the bus. The utilization of the control lines determines the type of transfer to be performed. As stated above, a MULTIBUS can support multiple masters; however, only one master may be in control of the bus at any given point in time.

4.4.1 MULTIBUS Power

MULTIBUS supports six supply voltages on the P1 bus. They are

1. +5 VDC—carried on eight pins
2. +12 VDC—carried on two pins
3. −12 VDC—carriedon two pins
4. +15 VDC—carried on two pins
5. −15 VDC—carried on two pins
6. Common—carried on eight pins

There are no provisions for on-board regulation of supply voltages. However, all supplies are tightly specified for line and load regulation. Line and load regulation for all supplies is specified at 1%. Peak-to-peak ripple for all power supplies is specified at not greater than 50 mV within a 50-MHz bandwidth. Transient response is defined as the time required for a supply to return to specification after a 50% change in load current and is specified as less than 500 µs. All MULTIBUS systems are specified to operate over a temperature range of 0°C (32°F) to 55°C (150°F). The humidity range is from 0 to 90% noncondensing. A summary of MULTIBUS power supply specifications is given in Table 4-1.

4.4.2 MULTIBUS Data Lines (DAT0/through DATF/)

MULTIBUS supports 16 bidirectional data lines, referred to as DAT0/ through DATF/. As the name indicates, the data lines carry the actual data to be transferred on each bus operation that involves a transfer of data. These lines are active low, meaning that a low voltage on a data bus line indicates the presence of a logic 1 level and a high voltage the presence of a logic 0. The data lines are always driven by tristate drivers and all data lines are terminated on the bus with 2.2-kΩ pull-up resistors to +5 VDC power supply. Both masters and slaves may drive data lines. A master will drive the data lines during a write data to slave operation or the slave will drive the data lines during a read data from slave operation.

There are three types of data transfer reads and writes supported by MULTIBUS. A 16-bit word can be transferred in one bus cycle using DAT0/ through DATF/, with DAT0/ being the least significant binary bit. The other two types of transfers make it possible to transfer a 16-bit value in two 8-bit bus cycles. The first cycle transfers data bits 0 through 7, called the even byte, on data lines DAT0/ through DAT7/. The second cycle transfers data bits 8 through 15, called the odd byte, on DAT0/ through DAT7/. A summary of data line transmitter and receiver requirements can be found in Table 4-2.

4.4.3 MULTIBUS Address Lines (A0/ through A23/)

The 24 active low address lines used by MULTIBUS give the bus an addressing range of 16 Mbytes of memory. The address lines are used to specify the location in memory of an I/O-mapped or memory-mapped device. All address lines are driven by tristate drivers because of the multiple master capability of MULTIBUS.

I/O devices may be either 8-bit or 16-bit I/O space-addressed. A master generating only an 8-bit I/O address is not required to specify the upper 8 bits of the 16-bit I/O address space, only the lower 8 bits carried on A0/

TABLE 4-1
MULTIBUS Power Logic Levels

Mnemonic	Voltage	Pins	Tolerance	Line and load regulation	Ripple
+5	+5 VDC	P1-3,4,5,6 P1-81,82,83,84	1%	0.3%	25 mV peak to peak
+12	+12 VDC	P1-7, 8	1%	0.3%	25 mV peak to peak
-12	-12 VDC	P1-79, 80	1%	0.3%	25 mV peak to peak
GND	Ground	P1-1,2,11,12 P1-75,76,85,86	Reference	Reference	Reference

Logic levels for active low signals			
State	Level	At receiver	At transmitter
0	High	2.00 to 5.25 V	2.40 to 5.25 V
1	Low	-0.5 to +0.8 V	0.00 to 0.50 V

Logic levels for active high signals			
State	Level	At receiver	At transmitter
1	High	2.00 to 5.25 V	2.40 to 5.25 V
0	Low	-0.5 to +0.8 V	0.00 to 0.50 V

through A7/. It is important for the designer of I/O-mapped devices to ensure that the device can be configured to respond to 8- or 16-bit I/O-mapped addresses. Only masters may drive address lines.

A summary of address line transmitters and receivers can be found in Table 4-3.

TABLE 4-2
MULTIBUS I Line: Termination and Drive Requirements

Mnemonic	Driver location	Driver type	Sink current (mA)	Source current (mA)	Output capacitance (pF)	Receiver location	Receiver low current (mA)	Receiver high current (mA)	Termination type	Termination value
INT0/-INT7/	SLAVE	OC	16	—	300	MASTER	-1.6	0.040	PULLUP	1K
DAT0/-DAT F/	MASTER OR SLAVE	TRI	16	-2	300	MASTER OR SLAVE	-0.8	0.125	PULLUP	2.2K
ADR0/-ADR19/	MASTER	TRI	16	-2	300	SLAVE	-0.8	0.125	PULLUP	2.2K
MRDC/, MWTC/	MASTER	TRI	32	-2	300	SLAVE	-2	0.125	PULLUP	1K
IORC/, IOWC/	MASTER	TRI	32	-2	300	SLAVE	-2	0.125	PULLUP	1K
XACK/	SLAVE	TRI	32	-0.4	300	MASTER	-2	0.125	PULLUP	510
INH1/, INH2	INHIBITING SLAVE	OC	16	—	300	INHIBITED SLAVES	-2	0.05	PULLUP	1K
BCLK/	1 PLACE	TTL	48	-3.0	300	MASTER	-2	0.125	PULLUP PULLDOWN	220 330
BREQ/	ALL MASTER	TTL	10	-0.2	300	ARBITER	-2	0.05	PULLUP	1K
BPRO/	ALL MASTER	TTL	3.2	-0.2	300	MASTER	-1.6	0.05	NONE	—
BPRN/	ARBITER	TTL	16	-0.4	300	MASTER	-4	0.10	NONE	—
LOCK/	MASTER	TRI	32	-2.0	300	ALL	-2	0.125	1 PLACE	1K
BUSY/, CBRQ/	MASTER	OC	30	—	300	MASTER	-2	0.05	1 PLACE	1K
INT/	MASTER	OC	32	—	300	ALL	-2	0.05	1 PLACE	2.2K
CCLK/	1 PLACE	TTL	48	-3.0	300	ANY	-2	0.125	PULLUP PULLDOWN	220 330
INTA/	MASTER	TRI	32	-2.0	300	SLAVE	-2	0.125	1 PLACE	1K

4.4 MULTIBUS System Bus

TABLE 4-3
MULTIBUS I P1 Connector Pin Assignment

| \multicolumn{3}{c|}{Component side} | \multicolumn{3}{c}{Clad side} |

Pin number	Mnemonic	Notes	Pin number	Mnemonic	Notes
1	GND		2	GND	
3	+5		4	+5	
5	+5		6	+5	
7	+12		8	+12	
9	....	Reserved in IEEE-796; old systems may be -5V	10	....	Reserved in IEEE-796; old systems may be -5V
11	GND		12	GND	
13	BCLK/		14	INIT/	
15	BPRN/		16	BPRO/	
17	BUSY/		18	BREQ/	
19	MRDC/		20	MWTC/	
21	IORC/		22	IOWC/	
23	XACK/		24	INH1/	
25	LOCK/		26	INH2/	
27	BHEN/		28	ADR10/	
29	CBRQ/		30	ADR11/	
31	CCLK/		32	ADR12/	
33	INTA/		34	ADR13/	
35	INT6/		36	INT7/	
37	INT4/		38	INT5/	
39	INT2/		40	INT3/	
41	INT0/		42	INT1/	
43	ADRE/		44	ADRF/	
45	ADRC/		46	ADRD/	
47	ADRA/		48	ADRB/	
49	ADR8/		50	ADR9/	
51	ADR6/		52	ADR7/	
53	ADR4/		54	ADR5/	
55	ADR2/		56	ADR3/	
57	ADR0/		58	ADR1/	
59	DATE/		60	DATF/	
61	DATC/		62	DATD/	
63	DATA/		64	DATB/	
65	DAT8/		66	DAT9/	
67	DAT6/		68	DAT7/	
69	DAT4/		70	DAT5/	
71	DAT2/		72	DAT3/	
73	DAT0/		74	DAT1/	
75	GND		76	GND	
77	....	IEEE-796 reserved	78	....	IEEE-796 reserved
79	-12		80	-12	
81	+5		82	+5	
83	+5		84	+5	
85	GND		86	GND	

4.4.4 MULTIBUS Control Bus

There are a number of lines that comprise the MULTIBUS control bus. A summary of transmitter and receiver requirements can be found in Table 4-2. I have grouped control signals into three general classes.

1. Utility and bus arbitration
2. Data transfer
3. Interrupt control

A brief description of the function of each of the signals is provided.

A general observation about MULTIBUS lines is that most lines are chosen to be active low lines. When a line is in a low voltage state, the line is asserted in the case of control lines or transmitting a logic 1 in the case of data and address lines. There are some exceptions to this general rule, however, so care must be taken during the design process.

Utility and Bus Arbitration Lines

Constant Clock (CCLK/) This TTL line is driven by only one source in each MULTIBUS system and is a general-purpose timing and control signal for any board in the system. It is a simple square wave signal with approximately a 50% duty cycle of a fixed frequency. The fixed frequency is determined by the specific hardware used but is typically close to 10 MHz.

Bus Clock (BCLK/) A 50% duty cycle TTL clock used to synchronize bus arbitration logic. Bus arbitration is the process by which two masters contend for control of the bus. The arbiter determines which device gains control of the bus. This signal may be changed in speed, single-stepped, or stopped, depending on requirements. It is generated by a single source. Proper design procedure indicates that every bus master should have the optional capability to drive the BCLK/ line. See the section on MULTIBUS I bus arbitration for more information on the use of this signal.

Initialize (INIT/) An open collector line driven by all masters and/or an external source such as a debounced front panel switch. It is usually asserted upon power-up or during system initialization, causing all of the boards in the system to be set to their initial condition.

Lock (LOCK/) This active low line is driven by the current bus master during read-modify-write (RMW) operations. It prevents access to dual port memory being used as a shared resource between processors between the read and write operations during the RMW operation.

Bus Request (BREQ/) An active low TTL signal driven by each master used for priority resolution during bus arbitration cycles. Unlike most lines on the bus, this line is not daisy-chained through all board slot

4.4 MULTIBUS System Bus

positions. It is used by the parallel priority resolution circuitry mounted on the mother board.

Bus Priority In (BPRN/)   This active low TTL non-daisy-chained signal is used by all bus masters during bus arbitration cycles. It can be driven by either the parallel priority resolution circuitry for systems using parallel priority resolution or by the next high-priority master in systems using serial priority resolution. An active low on this input to a master indicates to the master that it may begin the bus exchange process with the current master and gain control of the bus.

Bus Priority Out (BPRO/)   An active low TTL signal driven by all bus masters used for serial priority resolution. A master not contending for control of the bus will transparently send the same polarity out on its BPRO/ line as is received on its BPRN/ line input. A master contending for the bus will transmit an inactive high out of the BPRO/ line when an active low BPRN/ is received at its BPRN/ input. An active low on the BPRO/ line indicates to the next lower priority master that it may gain control of the bus.

Bus Busy (BUSY/)   This open collector active low signal is driven low by the current bus master to indicate that the bus is in use. It is used as a bidirectional handshake signal during bus exchange operations.

Common Bus Request (CBRQ/)   This open collector, active low line is driven by all bus masters. This signal is optional in that it is intended to maximize the performance of bus masters by indicating the presence or absence of a bus request to the current master.

Data Transfer Lines
- Memory Read Command (MRDC/)
- Memory Write Command (MWTC/)
- Input/Output Read Command (IORC/)
- Input/Output Write Command (IOWC/)

These four tristate command lines are driven by the current bus master, indicating the type of data transfer the bus master is requesting. An active low on either the IOWC/ or MWTC/ lines indicates that the bus master is performing a write operation. This low signal indicates to the receiving slave that the address and data bus lines all have stable data present. After the slave has taken the data, the command line returning high indicates the end of the transfer cycle to the slave. An active low on either the MRDC/ or IORC/ command lines indicates that the bus master is performing a read operation. A low signal on this line indicates to the transmitting slave that stable address data is present on the address bus lines. The command lines returning to the inactive high state indicate to the slave that the bus master has taken the data placed on the bus by the responding slave and that the slave can terminate the transfer process and release the bus data and control lines for another transfer.

Transfer Acknowledge (XACK/) This tristate line is driven by any responding slave in the system. It is used as a part of the data transfer handshake protocol with the master for all four type of data transfers, memory reads, memory writes, I/O reads, and I/O writes. The line is driven low by a slave responding to a data transfer request from the current bus master. An active low on this line indicates that the master's requested transfer has occurred, that is, data has been taken by the slave in the case of a write operation or data has been placed on the bus by the slave in the case of a read operation. This line is asserted by the slave until after the command line has been returned to inactive high by the bus master.

Inhibit (INH1/ and INH2/) These open collector lines can be driven by any (inhibiting) slave and are used by one slave to prevent bus activity by another slave during memory read or write operations. This feature allows overlaying of multiple devices in the same address space. This feature is convenient for the use of phantom ROMs or bootstrap loaders or for overlaying memory-mapped I/O into memory space.

Byte High Enable (BHEN/) This tristate line is driven by any bus master. It is a signal used as a vehicle to enable or disable 16-bit transfers on the data bus. It also adds the capability to use a combination of 16-bit masters with 8-bit slaves.

Interrupt Lines

MULTIBUS I Line Many busses have lines that fulfill multiple purposes. A given set of lines performs one function for one type of operation and a different function for another operation. This is the case for data lines. These lines are used to transfer data during data transfers but are also used to pass interrupt vector information during bus-vectored interrupts. Additionally, address lines ADR8/ through ADR10/ can be used to pass an interrupt code indicating which interrupt is being acknowledged by the master during bus-vectored interrupt acknowledge cycles.

Interrupt 0 through 7 (INT0/ through INT7/ These eight open-collector, active-low, prioritized-interrupt request lines can be driven by any slave. INT0/ has the highest priority, INT1/ the next highest, extending down to INT7/, which has the lowest priority.

Interrupt Acknowledge (INTA/) An active low, tristate signal driven by bus masters and used to acknowledge a slave's interrupt request. Two types of interrupts are supported by MULTIBUS. Bus-vectored interrupts make use of the INTA/ line while non-bus-vectored interrupts do not use this line. Bus-vectored interrupt acknowledged cycles can be generated as two-cycle INTA/ or three-cycle INTA/. A single 8-bit vector is passed to the master during two-cycle interrupt acknowledges and a full 16-bit interrupt address is placed on the data lines during a three-cycle interrupt acknowledge.

4.5 MULTIBUS Data Transfers

MULTIBUS supports four types of data transfers. I/O reads and writes and memory reads and writes are handled in a similar manner. All data transfers are originated by a bus master and responded by a slave. Note that the sense of all bus operations is always taken from the perspective of the bus master. A read operation is a read by the bus master or a transfer of data from a slave to the master. A write operation is a write by the bus master or a transfer of data from the master to the slave. Inhibit operations are a type of special data-transfer operation.

4.5.1 Data Write Operations

All write operations require the use of address lines, data lines, IOWC/ or MWTC/, and XACK/. Data, address, and command lines are all driven by the bus master, while XACK/ is driven by the slave. The IOWC/, MWTC/, IORC/, and IOWC/ lines are referred to as the "command" lines. The transfer sequence begins with the master asserting the address lines with the address of the slave which will take the data and placing the data to be transferred on the data lines. After the address lines have been stable for a minimum of 50 ns, the master asserts the IOWC/ in the case of a transfer to an I/O device or MWTC/ in the case of a transfer to memory. The addressed slave decodes the address, selecting the appropriate on-board device. When the slave detects the active low on the command line, the bus will have stable data. The slave loads the data into the selected on-board device. After the slave has captured the data, the slave asserts the XACK/ line by bringing the line low. This notifies the master that data has been taken by the slave. The master deasserts the command line and releases the address and data bus. The slave, detecting the inactive status of the command line, completes the transfer by releasing the XACK/ line. This process is illustrated in Figure 4-3.

4.5.2 Data Read Operations

Similar to writes, there are two read operations that may be performed on MULTIBUS: a read from I/O-mapped devices and a read from memory-

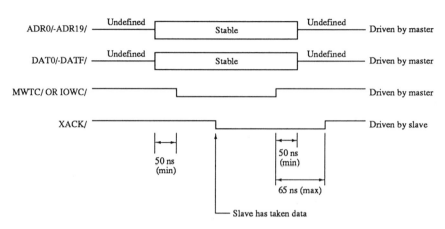

Figure 4-3 MULTIBUS memory or I/O write timing diagram.

mapped devices. Like all data transfer operations, the process begins with the master placing stable address data on the address lines. A minimum of 50 ns after the address data has become stable, the master asserts the IORC/ or MRDC/ "command" line, depending on the type of read being performed. The slave selects the device being addressed by the master. When the slave detects an active low on the command lines, the slave places the data to be read on the data bus and asserts the transfer acknowledge (XACK/) line. The master detects the asserted low on the XACK/ line which provides the signal from the slave to the master that stable data is on the data lines. The bus master loads the data into the on-board device needing the data and releases the command line to the inactive high state. The command line, being inactive, signals the slave that the data has been taken by the master. The slave responds by releasing the XACK/ line and setting the data lines to tristate mode, completing the read operation. This process is shown in Figure 4-4.

4.5.3 Bus Inhibit Operations

A bus inhibit operation occurs when two slaves occupy the same physical memory location. To prevent both slaves from responding to a read or write command, some protocol must be established so that one slave may inhibit the operation of the second slave. In this example, it will be assumed that ROM is overlaying a RAM location and that the ROM slave will inhibit the RAM when accessed. A read by the master is also assumed. Refer to Figure 4-5 for the timing diagram of the inhibit process.

4.5 MULTIBUS Data Transfers

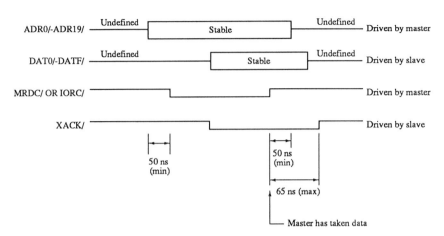

Figure 4-4 MULTIBUS memory or I/O read timing diagram.

The process begins as a normal master read operation with the master placing stable address information on the bus. The MRDC/ line is asserted by the master. Both the ROM and RAM slaves have decoded the address information, each selecting its respective on-board devices. Within 100 ns after the address data has become stable, the ROM slave asserts the INH1/ line. Note that INH1/ is used by ROM to inhibit RAM and INH2/ is used by RAM to inhibit ROM. The RAM slave detects the active INH1/ line and

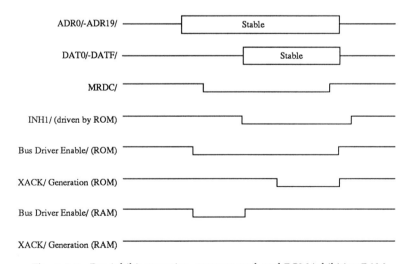

Figure 4-5 Bus inhibit operation, memory read, and ROM inhibiting RAM.

immediately deselects the on-board device and disables any bus drivers that may have been briefly selected. The ROM slave then places data on the data lines and asserts the XACK/, completing the data read in the normal manner in which a memory read occurs.

4.6 Multiple Masters

A single MULTIBUS chassis can support a number of masters. There is no theoretical limit to the number of masters but as a matter of practicality it seldom exceeds more than five to eight. A large number of masters can result in very poor overall system speed performance without very careful system design. As in all bus systems, there can be only one bus master at any given point in time. If another master wishes to gain control of the bus, a two-step process must be undertaken. The first step is called bus arbitration. It is the process by which the bus determines which of several masters that may be requesting the use of the bus will be granted ownership or control. It may be that none of the masters requesting the bus will be granted control but the current master will retain control of the bus. The second step in the process is called the bus exchange. This process defines an orderly methodology by which the current bus master can relinquish control of the bus and the new master gain control.

4.6.1 Bus Arbitration

Bus arbitration makes use of the following lines:

1. BPRN/
2. BPRO/
3. BREQ/
4. CBRQ/

There are two techniques by which bus arbitration is implemented on MULTIBUS. The serial technique is the simplest to implement, making use of the BPRN/ and BPRO/ lines. This technique, also called daisy chain, suffers from also being somewhat slow, limiting the number of masters that can be supported to less than four. The parallel technique is slightly more complex; however, it is significantly faster and does not limit the number of bus masters. This technique uses the BPRN/ and BREQ/ lines along with some logic that must be installed on the backplane.

Serial Prioritization

Serial prioritization is also called daisy-chain prioritization. It is accomplished using the BPRO/ and BPRN/ lines. Recall that these lines are

4.6 Multiple Masters

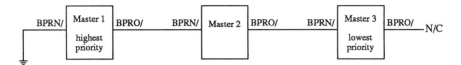

Figure 4-6 MULTIBUS I serial master prioritization.

not bussed to each slot position. The serial priority configuration is shown in Figure 4-6. The highest priority master has its BPRN/ line jumpered to ground, always giving that master control of the bus when needed. When using the bus, the BPRO/ line is kept inactive high, indicating to the next lower priority master that the master may not gain control of the bus even if it has a bus request pending. Each successive lower master generates a high if its corresponding BPRN/ is high from the next higher priority master.

When the highest priority master no longer needs control of the bus, the BPRO/ line is made active low, indicating to the lower master that the bus is available for use. If the priority two master does not need the bus, it ripples the low on to the next lower priority master. The ripple continues until a master needing the bus receives the low on its BPRN/ input. It does not ripple the low to its BPRO/ output but instead uses the bus exchange process and gains control of the bus. If the BPRN/ line at the low-priority master goes low while it is in control of the bus, it is a signal from a higher priority master that it wants control of the bus and the low-priority master must begin the master exchange protocol to relinquish control of the bus. The entire ripple process through all masters must take place within one BCLK/ period.

Parallel Prioritization

The parallel priority technique makes use of the BREQ/, BPRN/, and CBRQ/ lines and bus arbitration circuitry located on the backplane of the mother board. The two devices commonly used are a 74148 priority encoder and a 74138 decoder. Each master desiring use of the bus generates a BREQ/ low signal that is routed to the 74148 inputs. The 148 device has eight inputs and three outputs. The outputs are binary coded signals corresponding to the highest priority input. For example, if two masters are requesting the bus and the highest priority master BREQ/ is routed to input 7 and the lower priority master requesting the bus is routed to input 2, the 74148 will generate a 7 or binary 111b at its outputs. The decoder has three inputs and eight outputs. It generates an active low output on the line the input selects. The decoder will then take the 111b at its input and generate an active low on output 7. Output 7 is routed to the BPRN/ line of the highest priority master.

Each master desiring the use of the bus will generate a BREQ/ and then begin monitoring the BPRN/ line. When the line goes low, the bus exchange process may be started by the master. If a master is in control of the bus and its BPRN/ line becomes inactive high, the master must complete the current instruction and begin the bus exchange process, losing control of the bus. This process is shown in Figure 4-7. The parallel arbitration process illustrated here using the priority encoder and decoder is a simple hierarchial prioritization scheme. It is possible to implement more complex types of prioritization schemes, such as round robin, using different logic.

The CBRQ/ line is used to save bus exchange overhead or speed up the exchange process. If a master in control of the bus does not sense a low on the open collector CBRQ/ line, no other master is requesting the bus. If the CBRQ/ line is low, there is some master other than the master currently in control of the bus requesting the bus. It may be either a higher or lower priority master. If it is a higher priority master it will lose the bus when the BPRN/ line at its input goes high. If a lower priority master is requesting the bus, the master in control of the bus has the option to relinquish the bus or continue in control. Using CBRQ/ can save time during bus exchange operations also. In systems in which masters change rapidly, this may result in improved system speed.

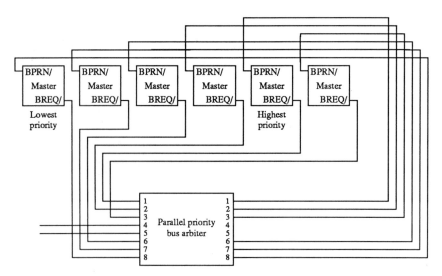

Figure 4-7 MULTIBUS I parallel master prioritization.

4.6.2 Bus Exchange

The bus exchange process is similar whether the serial or parallel prioritization technique is used. The lines used in the bus exchange process are

1. BCLK/
2. BREQ/ (BPRO/)
3. BPRN/
4. BUSY/

The parallel scheme will be discussed here. The process is shown in Figure 4-8 (lower half).

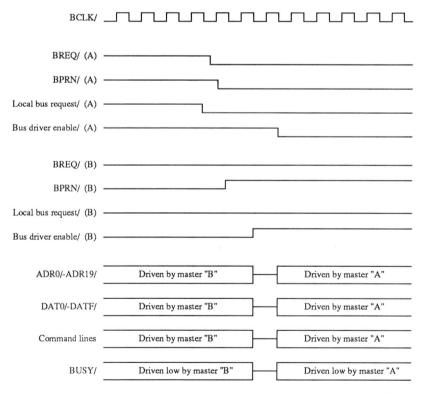

Figure 4-8 MULTIBUS I bus arbitration and exchange operation, parallel priority resolution, and master "A" gaining control from master "B."

All bus exchange signals are synchronous with the falling edge of the BCLK/ signal. When the exchange process begins, assume that master "A" is a low(er) priority master and is in control of the bus, while master "B" is a high(er) priority master and will request use of the bus. Master "B" begins the process by making its BREQ/ line active low. The arbiter determines that a higher priority master has requested the bus from a lower priority master and makes the BPRN/ line inactive high on master "A" and active low on master "B." This is the arbitration phase and is completed in less than 1 BCLK/ cycle. Master "A" still wants the bus but must relinquish the bus to "B." However, it continues to assert BREQ/. It will regain control of the bus when master "B" is completed, provided there are no other higher priority requests pending.

Master "A" completes the instruction it is currently executing and then tristates the command, address, and data lines. Master "A" allows the open collector BUSY/ line to go high, indicating to master "B" that the bus has been released by master "A." Master "B," sensing that master "A" has completed its last bus operation because the BUSY/ line indicates the bus is no longer busy, drives the BUSY/ line low, indicating that it has control of the bus and enables its command, address, and data line drivers and begins performing bus operations.

4.7 Interrupts

There are two general methodologies employed by MULTIBUS to handle bus interrupts. They are bus-vectored and non-bus-vectored interrupts. Non-bus-vectored interrupts do not pass interrupt vector address information over the bus, while bus-vectored interrupts do pass interrupt vector addresses over the bus. When using non-bus-vectored interrupts, the interrupt vector address is generated by an interrupt controller on the bus master responding to the interrupt and is passed to the processor over a local bus on the master.

4.7.1 Bus-Vectored Interrupts

Bus-vectored interrupts make use of the following lines:

1. INT0/ through INT7/
2. INTA/
3. XACK/
4. ADR8/, ADR9, and ADR10/
5. DAT0/ through DAT7/

4.7 Interrupts

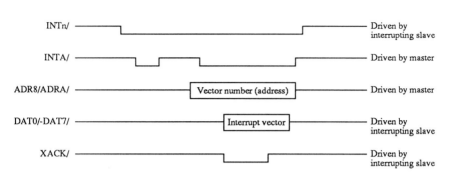

Figure 4-9 MULTIBUS I interrupt timing diagram: two-cycle bus-vectored interrupt.

There are two types of bus-vectored interrupts that may be supported by MULTIBUS. Two-cycle and three-cycle interrupts may be generated. Two-cycle interrupts pass an 8-bit vector over the bus to the master, while three-cycle interrupts pass a 16-bit vector over the bus to the master. The timing diagram for a two-cycle bus-vectored interrupt is shown in Figure 4-9.

The process begins when a slave generates an interrupt on one of the INTn/ lines. The master recognizes the interrupt by generating the first INTA/ cycle which is not used by the slave. The master generates a second INTA/ cycle, placing the priority of the INTn/ line onto the address bus lines ADR8–ADR10/. For example, if INT3/ is being recognized by the master, ADR8/–ADR10/ will contain a binary 3. The slave driving the INTn/ line low recognizes the INTA/ and decodes ADR8/–ADR10/ to checking to ensure that the master is responding to the interrupt it is asserting no anther interrupt. If the master is responding to the correct INTn/ line, the interrupting slave drives valid vector information onto DAT0/ through DAT7/ and generates an active low on XACK/. The master detects the low on XACK/ and captures the vector on the data bus and releases the INTA/ line to its inactive high state.

4.7.2 Non-Bus-Vectored Interrupts

Non-bus-vectored interrupts are simpler for both the slave and master. However, usually a programmable interrupt controller is used by the master and this must be programmed upon system initialization for correct operation. The interrupting slave generates an interrupt on one of the INT0/ through INT7/ lines. The master recognizes the interrupt request branching to the interrupt service routine. Data transfers are performed over the bus, causing the interrupting slave to release the interrupt line.

5

MULTIBUS II

5.0 MULTIBUS II Overview

MULTIBUS II, along with VMEbus discussed in the following chapter, is a popular and powerful bus commonly used in new designs. Nearly every major processor in the commercial marketplace has been designed into a MULTIBUS II board. It continues to grow in popularity, with a large number of new designs currently being released.

MULTIBUS II was conceived by Intel Corporation and was designed as the natural successor to MULTIBUS I, extending the architecture from the 16 bits of MULTIBUS I to 32 bits. It was intended to be an open-architecture bus that was processor-independent and completely standardized. The development began in the early 1980s with the formation of a consortium of companies that would develop the standard. It is intended for high-performance, multiprocessor applications. The bus is optimized for block transfer of data with transfer speeds of up to 40 Mbytes/s. The bus is actually more of a network connection scheme for a multiprocessing network. Boards in MULTIBUS II tend to be intelligent processing nodes with each running in a coupled manner with the other boards (or processing nodes) in the system.

MULTIBUS II was originally part of the Future Bus (P896) standard in 1983 but separated in 1986 to become the P1296 working group. The final specification is ANSI/IEEE Std 1296 and was released in 1987.

Like VMEbus, MULTIBUS II is a complex bus capable of supporting many types of operations. This chapter is intended as an introduction to MULTIBUS II and is not exhaustive in its treatment.

5.1 Basic Features and Capabilities of MULTIBUS II

Some important features supported by MULTIBUS II include

1. Multiple masters
2. Block data transfers to 40 Mbytes/s
3. 32-bit multiplexed address and data bus
4. Dynamic bus sizing permitting 8-, 16-, 24-, and 32-bit transfers
5. Several standard board sizes
6. Bus exception (error) handling
7. Synchronous bus transfers using a 10-MHz transfer rate
8. Parity checking on each transfer

MULTIBUS II has several subgroup busses, each containing a collection of lines that perform a specific set of functions. Some lines perform multiple functions, depending on the type of bus operation being performed.

The major subgroups of busses are

1. Central control signal bus
2. Arbitration bus
3. Address/data bus
4. Exception handling bus
5. System control signal bus

5.2 MULTIBUS II Mechanical Specifications

MULTIBUS II is designed to be mechanically consistent with the requirements of ANSI/IEEE Std 1101-1987. The complete specification details the requirements for card cages, backplanes, boards, and front panels. MULTIBUS II permits two common sizes of boards. The double-height board basic dimensions are shown in Figure 5-1. It uses two 96-pin connectors. Note that all of the MULTIBUS II parallel system bus (PSB) is implemented on P1. P2 is used for some power connections, with the rest being used for user-defined functions. The triple-height board uses three 96-pin connectors. Note again that P2 and P3 may contain user-defined functions. The basic dimensions of the double-height 6U board are 9.19 by 8.66 in. The triple-height 9U board is 14.44 by 8.66 in. A deeper modified triple-height board is 14.44 by 11.02 in. Other board sizes (less frequently seen) are also supported by the ANSI/IEEE Std 1101

5.2 MULTIBUS II Mechanical Specifications

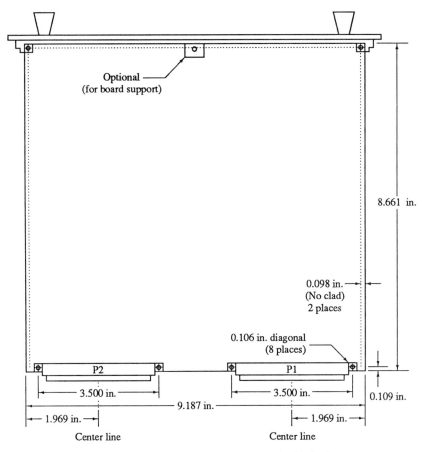

Figure 5-1 MULTIBUS II board dimensions: double high (6U).

specification and may be used to implement MULTIBUS II designs. The 3U small-board standard is not supported by MULTIBUS II. The triple-height board outline is shown in Figure 5-2. All connectors are 96-pin (3 rows of 32 pins) connectors on a 0.1 in. grid.

In MULTIBUS II card cages with 12 or fewer slots, card slot 0 is on the left as viewed from the front, with card slot numbers increasing moving toward the right. For card cages with 13 through 21 card slots, card slot 0 is in the center of the card cage for card cages with an odd number of slots or one right of center with an even number of slots. Numbers increase moving toward the right to the right-most slot. The left-most slot is one higher than the right-most slot number and increases moving to the right until slot 0 is reached.

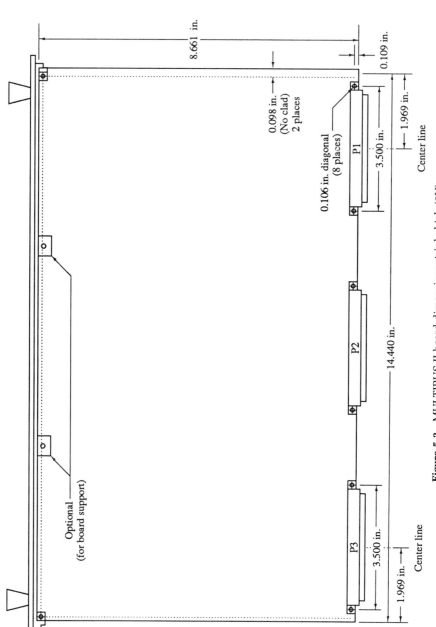

Figure 5-2 MULTIBUS II board dimensions: triple high (9U).

5.2.1 Pin Assignments

The double-height board has 192 pins on two 96-pin connectors. Unused pins on P2 and P3 on the triple-height board are available for individual user's definition. All connectors are identical and have three rows of 32 pins. The rows are labeled A, B, and C, while the pins in each row are numbered. Individual pins are given alphanumeric designators such as C12, indicating row C, pin 12. The complete MULTIBUS II pin assignments for both P1 and the optional P2 are shown in Figure 5-3.

5.3 MULTIBUS II Operations

MULTIBUS II has three types of boards or "modules." There are requesting agents, replying agents, and both requesting and replying agents. Requesting agents may be thought of as bus masters and may have control or "ownership" of the bus, during which they initiate bus operations. Replying agents may be thought of as bus slaves and reply to operations initiated by requesting agents. Requesting and replying agents are dual-mode boards.

There are only three basic types of operations that are performed on MULTIBUS II. They are

1. Arbitration operations
2. Exception handling operations
3. Data transfer operations

Note that the bus has the capability to perform any of these operations concurrently. Data transfer operations can be further divided into read data and write data transfer operations. Any data transfer operation is divided into two cycles. During the first cycle, the device wishing to perform a data transfer must arbitrate for control or ownership of the bus. Once control of the bus is obtained, the device may then perform the data transfer(s). Transfer of data may be any number of bus transfers, including one. Any operation can be terminated by any device detecting an error or exception condition.

5.3.1 Transfer Operations

The module in control or owning the bus initiates all data transfers. The control lines are used to send the type of transfer or command, handshake, and end of transfer (EOT). All transfers are divided into a request phase and a transfer phase. During the request phase, the current

	Connector P1				Connector P2		
Pin number	Row A	Row B	Row C	Pin number	Row A	Row B	Row C
1	Ground	PROT/	Ground	1	Ground		Ground
2	+5	DCLOW/	+5	2	+5		+5
3	+12	+5 BATT	+12	3			
4	Ground	SDA	BCLK/	4			
5	TIMEOUT/	SDB	Ground	5			
6	IDLACH/	Ground	CCLK/	6			
7	AD0/	AD1/	Ground	7			
8	AD2/	Ground	AD3/	8			
9	AD4/	AD5/	AD6/	9			
10	AD7/	+5	PAR0/	10			
11	AD8/	AD9/	AD10/	11			
12	AD11/	+5	AD12/	12			
13	AD13/	AD14/	AD15/	13			
14	PAR1	Ground	AD16/	14			
15	AD17/	AD18/	AD19/	15			
16	AD20/	Ground	AD21/	16			
17	AD22/	AD23/	PAR2/	17			
18	AD24/	Ground	AD25/	18			
19	AD26/	AD27/	AD28/	19			
20	AD29/	Ground	AD30/	20			
21	AD31/	Reserved	PAR3/	21			
22	+5	+5	Reserved	22			
23	BREQ/	RESET/	BUSERR/	23			
24	ARB5/	+5	ARB4/	24			
25	ARB3/	RSTNC/	ARB2/	25			
26	ARB1/	Ground	ARB0/	26			
27	SC9/	SC8/	SC7/	27			
28	SC6/	Ground	SC5/	28			
29	SC4/	SC3/	SC2/	29			
30	-12	+5 BATT	-12	30			
31	+5	SC1/	+5	31	+5		+5
32	Ground	SC0/	Ground	32	Ground		Ground

Figure 5-3 MULTIBUS II P1 and P2 pin assignments. Note: (1) Pins P1 B4 and P1 B5 (SDA and SDB) are for future serial bus, and (2) blank pins are user defined.

bus owner or requesting agent has acquired bus ownership by means of an arbitration operation. It places command data on the control lines and address data on the address/data lines. The command broadcasts the type of transfer that is to be performed. The reply phase can occupy one or more bus cycles, depending upon the number of data transfers that are to occur. If there is only one transfer, the reply phase consists of placing data on the address/data lines and generating a handshake and operation on the control lines. If there are multiple data transfers, the reply phase consists of placing a sequence of data on the address/data lines and generating a string of handshakes, one for each transfer, followed by a handshake and EOT on the last transfer. Data is driven onto the address/data bus by the requesting agent (master) during write operations and driven onto the data bus by the replying agent (slave) for read operations. At time flowchart for a transfer operation is shown in Figure 5-4.

5.3.2 Exception Operations

The purpose of exception operations is to report the occurrence of an error and to provide a methodology for recovering from the consequences of an error. All modules perform an exception operation in the event that any module detects an error. The exception operation consists of a signal phase and a recovery phase. Both transfer and arbitration operations are stopped when an exception occurs. During an exception operation, the exception being performed is in response to an error occurring during the preceding transfer.

5.3.3 Arbitration Operations

As in all busses, only one master may be in control of the bus at any given time. The arbitration operation is the method by which the masters gain exclusive control or ownership of the bus. As with the transfer and exception operation, the arbitration operation is performed in two phases. The resolution phase occurs first and is used to determine which will be the next master or requester agent to have control of the bus. The acquisition phase is the process by which the current master relinquishes control of the bus to the new master.

The methodology by which the arbitration is performed is a simple fixed priority scheme. The priority of each master is binary encoded and loaded into an on-board register. The priority is established during a power-up reset during which a central services module (CSM) loads priority data into each master. The process does ensure that all masters, even those with low priority, do receive bus ownership, however. All bus

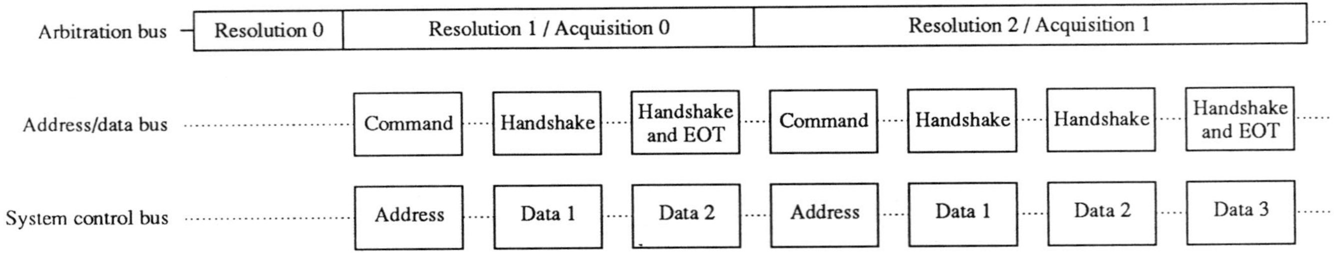

Figure 5-4 MULTIBUS II transfer operation and time flow chart.

masters enter the arbitration process together. No new master may contend for the bus until every master in the current group requesting the bus has received ownership of it. Note also that when a master has control of the bus, that master will be performing transfers over the bus. While those transfers are in progress, the arbitration for the next master is being performed. The EOT command signals the end of ownership by the current master. The new master has already been decided and then acquires the bus. This process may be thought of as a one-deep master queue. The new master is always ready to acquire the bus even before the old master has completed its transfer operations. This process is shown in Figure 5-5.

5.4 MULTIBUS II Line Descriptions

All operations on MULTIBUS II are synchronous. All lines are to be sampled on the falling edge of the system clock or CCLK/.

All bus signal lines are placed in one of the subgroup busses outlined in Section 5.1. The arbitration operations group contains 7 lines, system control group 10 lines, exception operation group 2 lines, address/data group 36 lines, and central control group 7 lines, for a total of 62 lines, not including power and ground.

5.4.1 Address/Data Group

AD31/–AD0/ These 32 active low lines perform the function of carrying both address and data. They are used to transfer address information during the request phase of a data transfer and data during the reply phase. Data are driven onto the AD/ lines by the bus owner during a write operation and driven onto the AD/ lines by the responding selected replying agent or slave during a read operation. MULTIBUS II supports 8-, 16-, 24-, and 32-bit transfers so not all AD/ lines may be used for all transfers.

PAR3/–PAR0/ These four active low lines are used to indicate the parity of each byte on AD0/ through AD31/ during a data transfer operation. PAR0/ is used to indicate the parity of the least significant byte placed on AD0/ through AD7/, while PAR1/ indicates the parity on AD8/ through AD15/, PAR2 the parity on AD16/ through AD23/, and PAR3/ the parity on AD24/ through AD31. The parity used by MULTIBUS II is even, meaning that the parity bit is set in order to make the total number of asserted bits, including the parity bit in any given byte, an even number. Those parity bits used for bytes not driven by a device using less than the four bytes available on the bus are not driven valid.

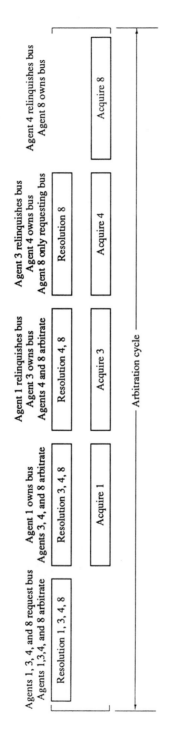

Figure 5-5 MULTIBUS II arbitration operation and time flow chart.

5.4 MULTIBUS II Line Descriptions

Line	Function performed
SC 0/	L = request
SC 1/	L = lock
SC 2/	Data transfer width bit 0
SC 3/	Data transfer width bit 1
SC 4/	Address space identifier bit 0
SC 5/	Address space identifier bit 1
SC 6/	L = data write, H = data read
SC 7/	Unused, driven high
SC 8/	Even parity for SC4/ through SC7/
SC 9/	Even parity for SC0/ through SC3/

Figure 5-6 MULTIBUS II request phase function and system control lines.

5.4.2 System Control Group

SC9/-SC0/ These active low system control signals are used to define and control the types of transfers between bus devices. During the request phase, the bus owner defines the system control lines and during the reply phase the bus owner drives SC0/, SC1/, SC2/, SC3/, and SC9/. The selected replying agent drives the remaining lines. The function of each line during the request phase is shown in Figure 5-6. The function of each line during the reply phase is shown in Figure 5-7.

SC9/ This is the parity bit for SC0/ through SC3/ during both request and reply. Even parity is the convention used, meaning that the number of asserted bits on SC0/-SC3/ and SC9/is always even.

Line	Function performed
SC 0/	H = reply phase
SC 1/	L = lock
SC 2/	L = EOT, H = Not EOT
SC 3/	L = requesting agent ready for transfer
SC 4/	L = replying agent ready for transfer
SC 5/	Replying agent status, bit 0
SC 6/	Replying agent status, bit 1
SC 7/	Replying agent status, bit 2
SC 8/	Even parity for SC4/ through SC7/
SC 9/	Even parity for SC0/ through SC3/

Figure 5-7 MULTIBUS II reply phase function and system control lines.

SC8/ This is the parity bit for SC4/ through SC7/ during both the request and reply phases. As in the case of the SC9/ parity bit, even parity is used.

SC7/–SC5/ During the reply phase, these bits represent the status of the replying module or agent. The defined codes are

SC7/ SC6/ SC5/	Code definition
000	Not used/reserved
001	Not used/reserved
010	Data error
011	Resource not available
100	Multiple errors
101	Continuation error
110	Data transfer width error
111	Normal completion, no errors

A continuation error occurs when a replying agent is unable to continue after the transfer is completed. Multiple errors is the "catch all" code used when the transfer is badly botched and the replying agent cannot make any sense of the requesting agent. A data transfer width error occurs when, for example, a 16-bit replying agent is required to perform a 32-bit transfer.

SC4/–SC5/ During the request phase, these bits are used to indicate the address space for which the transfer operation is to take place. MULTIBUS II supports four address spaces. They are memory, I/O, interconnect, and message space. The definition of address space is

SC5/ SC4/	Address space selected
00	Interconnect
01	Message
10	I/O
11	Memory

During the reply phase, SC5/ is one bit of the three-bit agent status, as discussed above. SC4/ is used during the reply phase by the agent to indicate whether the agent is ready to perform the transfer. When the line is active low, the agent is ready to perform a read or write operation. When high, the agent is not ready. Note that SC2/ and SC3/ indicate the width of data transfer during the request phase. The transfer width definition is

5.4 MULTIBUS II Line Descriptions

SC3/ SC2/	Transfer width
00	32 bits
01	24 bits
10	16 bits
11	8 bits

SC1/ This line performs the same function during both the request and reply phases. It is driven by the bus master (owner). When driven low, the bus resource is locked in place by the current master. All requests for change of ownership of the bus are placed on hold and the current bus owner retains ownership of the bus. If multiport devices are used in the system, the lock also prevents any other system master from gaining access to these resources. All bus resources are dedicated to the current bus owner until the SC1/ is made inactive.

SC0/ This line also retains the same function during both the request and reply phases and is used to indicate the phase of the transfer. When SC0/ is driven active low, the current transfer is in the request phase. When SC0/ is driven high, the current transfer is in the reply phase.

5.4.3 Exception Operation Group

TIMEOUT/ This active low signal indicates one of two error conditions. The central services module (CSM) drives TIMEOUT/ and indicates that either a replying agent took too long to complete a transfer or a bus master has owned the bus for an excessive length of time. The point at which the time becomes excessive is user-defined.

BUSERR/ This active low line is asserted by any agent and indicates a data integrity problem on the bus. This can be a parity error on the AD31/–AD0/ lines or the SC/ lines.

5.4.4 Arbitration Signal Group

ARB5/–ARB0/ Depending on the function being performed by the bus, these active low, open collector lines perform three services. During a reset, two functions are performed. They assign card slot identification data and arbitration priority data to each board during a reset. During normal operation they carry arbitration data.

ARB5/ carries a special function during reset. When driven low or active by the CSM, card slot identification data is driven onto ARB0/ through ARB4/. When ARB5/ is high during a reset, arbitration priority data is being placed on ARB4/ through ARB0/. All agents must load card slot identification and all masters must load arbitration priority data during a reset.

During normal operation, arbitration data is placed on these lines. The resolution of bus ownership starts during the resolution phase when all agents requesting use of the bus place their arbitration priority onto the ARB/ lines. These lines are bitwise ORed. The requesting agent with the arbitration priority matching its assigned priority will be the next agent to own the bus.

The ARB5/ line is not normally used for arbitration; however, it may be used for a high-priority request to bus ownership. If more than one requesting agent requests the use of the bus by means of asserting ARB5/, the conflict is resolved by using the same wired or technique on ARB4/ through ARB0/ discussed above.

BREQ/ The active low bus request line is used to ensure a fair arbitration scheme in which even the lowest priority masters will have access to the bus. All masters requiring ownership of the bus must assert their respective open collector BREQ/ lines. As each master sequentially receives and relinquishes control of the bus, that master releases the BREQ/ line on that board. After the last (lowest priority) master relinquishes control of the bus, the BREQ/ line becomes inactive high, indicating that the current arbitration cycle is complete and all requesting agents requiring use of the bus have had ownership. Once the line is high (inactive), bus masters are free to again assert BREQ/ on their respective boards and begin the arbitration cycle again.

5.4.5 Central Control Group

BCLK/ The bus clock is driven by the CSM and is used to synchronize all bus operations. Data is driven and sampled on the falling edge of this clock. It generally operates at a fixed frequency of 10 MHz. However, all boards must be designed to operate with a BCLK/ frequency of anywhere from DC to 10 MHz. It may be operated at very low frequencies to single-step the system during debugging.

CCLK/ The constant clock signal is driven by the CSM and operates at twice the frequency of the BCLK/ signal. It is used for general-purpose timing.

IDLACH/ The active low ID latch signal is used only during a reset. It is driven by the CSM and is used by the boards to latch either the arbitration priority data or the card slot ID number. The level of ARB5/ is used to indicate which is to be loaded. A module "knows" when to accept arbitration data or card slot ID by monitoring the status of AD20/ through

AD1/. The CSM will drive one of the AD/ lines active for the device to which the arbitration data or card slot ID is being sent for each slot. For example, AD5/ will be driven low when the CSM is sending arbitration and card slot ID to the module installed in card slot number 5.

RESET/ The active low reset signal is driven by the CSM and causes the system to be set to its initialized state. Reset is one of three type of initialization. There are a warm start, cold start, and power failure restart. See also the DCLOW/ and PROT/ signals.

PROT/ The protect signal is driven by the CSM. An active low on this line indicates that system power supplies or a power supply is out of specification. Battery back-up devices use this signal to prevent access to system-critical information that must not be corrupted during a power failure. This signal is not synchronous with the BCLK/ signal.

DCLOW/ The DC power low signal is used to signal devices in the system that a power failure is imminent. The active low signal is also asynchronous with the BCLK/ signal. It gives time to system resources to store or update critical data in battery-backed areas or memory.

RSTNC/ The open collector, active low reset not complete is driven low by any device requiring an extended period of time to complete a reset operation. No device may perform any bus operations as long as RSTNC/ is driven active and it must be released to inactive status as soon as possible by all agents after a reset operation. All bus agents must monitor the status of RSTNC/.

5.5 Timing Requirements for Bus Operations

5.5.1 Reset Operation

During a reset operation, the card slot ID and arbitration priority code is assigned and loaded into each agent or board in the system. The central services module (CSM) is always installed in card slot 0 so the priority and card slot ID is never assigned to slot 0. The timing diagram for the loading of arbitration priority codes and card slot IDs is shown in Figure 5-8.

5.5.2 Transfer Operations

All transfer operations are initiated by the current bus master or owner of the bus. Bus ownership is determined by the arbitration operation. Two groups of signals are used in the transfer operation: the address data (AD) lines and the system control (SC) lines. There are two of many

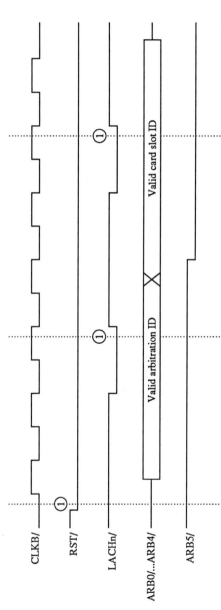

Figure 5-8 Card slot and arbitration priority reset timing. ① may insert additional clock cycles here.

5.5 Timing Requirements for Bus Operations

different types of transfers illustrated here. The first is a simple, single, 24-bit write transfer to a replying agent from a requesting agent. This is shown in Figure 5-9. The second is a two-transfer, 32-bit read transfer. This is shown in Figure 5-10.

The replying agent determines that the last transfer is in process by examining the status of SC2/. The requesting agent will assert SC2/ during the last transfer.

Broadcast Messages

A special type of transfer is the broadcast message intended for all agents and is targeted to occur in message space only. During a broadcast message, the replying agents do not provide status information at the end

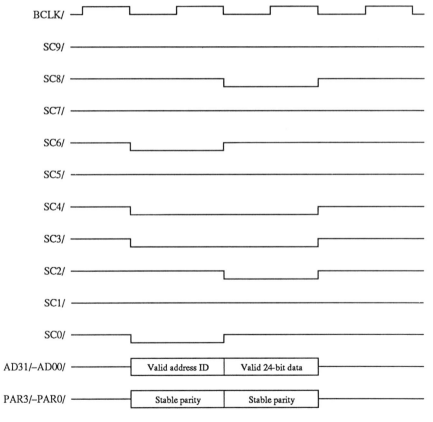

Figure 5-9 MULTIBUS II single 24-bit write transfer and I/O space transfer.

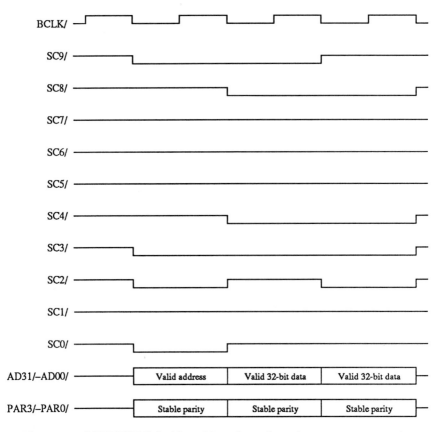

Figure 5-10 MULTIBUS II double 32-bit read transfer and memory space transfer.

of transfer. Since message space is only 256 bytes, AD7/ through AD0/ are driven valid during a broadcast message.

Types of Address Space

MULTIBUS II supports four types of data transfers, with the destination of the transfer being determined by the state of SC5/ and SC4/ during the request phase of a transfer. The four memory maps are

1. Interconnect space
2. Message space
3. Memory space
4. I/O space

Interconnect space is a special area of memory set aside for diagnostic and initialization purposes. It is accessed by the card slot address and is

5.5 Timing Requirements for Bus Operations 77

limited to 8-bit read/write transfers only. Broadcast and sequential transfers into interconnect space are not supported. The complete address during an interconnect transfer is only 16 bits. AD15/ through AD11/ (5 bits) are used to identify the card slot to which the transfer is to take place. AD1/ and AD0/ are not driven. AD2/ through AD10/ (9 bits) is used for the register address on the selected board. This gives a total interconnect addressing range of 512 bytes on each board.

Message space allows for the passing of data or messages between modules in a system. All types of data transfers are supported, including 8-, 16-, 24-, and 32-bit transfers that can be both single and block types of transfers. However, message address is not auto-incremented on block message space operations. Broadcast messages into message space are also supported. The address is 16 bits. AD15/ through AD8/ contains the source address and AD7/ through AD0/ contains the destination address. Note that message passing is unidirectional. It is a write operation only. Data may not be read from message space over the bus.

Memory space supports the broadest types of transfers over the greatest range of functions; 8-, 16-, 24-, and 32-bit read/write transfers are supported. Both single and block transfers are supported. The total addressing range is 32 bits or 4 gigabytes. During block transfers, the replying agent auto-increments the address during subsequent sequential transfers for block operations. The auto-increment technique depends upon the type of block transfer being performed. Note that MULTIBUS II does not support 24-bit block transfers or unaligned 16-bit block transfers. An unaligned word transfer is a word (16-bit) transfer into an odd byte address. During 8-bit or byte block transfers, the byte address of the replying agent increments by one (byte address) after each transfer. During 16-bit block transfers, the byte address of the replying agent increments by two (byte addresses) after each transfer. During 32-bit block transfers, the byte address of the replying agent increments by four (byte addresses) after each transfer.

Input/output space is used to perform transfers to I/O-mapped devices. The I/O address space is 16 bits, allowing 64k I/O ports in the system. I/O transfers can be 8, 16, 24, or 32 bits with both single and block read/write transfers being supported. Block transfers of 24 bits or unaligned 16-bit transfers are not supported. There is no auto-address increment feature for I/O address space, so sequential transfers during a block operation are performed on the same physical port.

5.5.3 Exception Operations

Exception handling is used to report the occurrence of an error condition and provide a mechanism for recovery from such an event. As mentioned earlier, there are timeout and bus error exceptions. A timeout error is caused by an agent requiring too much time to complete a given bus transaction. Bus errors are caused by conditions such as incorrect width, illegal control signals, and parity errors. Error detection and correction are handled in two phases. The first is the signal phase, during which the existence of an error is signaled to all bus agents, followed by a bus idle time called the recovery phase.

When the bus master detects a signaled exception, it is required to terminate bus activity. The bus is held in this inactive state until the TIMEOUT/ or BUSERR/ signals are both returned to their inactive state. This event marks the end of the signal phase. The signal phase must last at least one BCLK/ period, but there is no upper limit to the number of BCLK/ periods it may last. As soon as the TIMEOUT/ or BUSERR/ signals

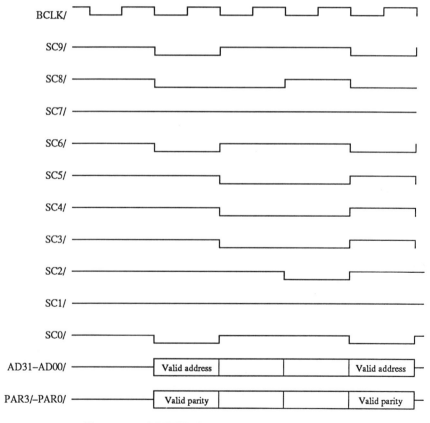

Figure 5-11 MULTIBUS II exception operation timing.

Line or line group	Type of line	Driver requirements		Receiver requirements		Termination requirements		Notes
		Ih	Il	Ih	Il	Pull up	Pull down	
Address/data group								
AD31/-AD0/	Tri-state	3	48	0.1	1	330	470	Terminate both ends of backplane
PAR3/-PAR0/	Tri-state	3	48	0.1	1	330	470	Terminate both ends of backplane
System control group								
SC9/-SC0/	Tri-state	3	64	0.1	1	220	330	Terminate both ends of backplane
Exception group								
TIMEOUT/	TTL	3	48	0.1	1	330	470	Terminate both ends of backplane
BUSERR/	Open collector		60	0.1	0.9	220	330	Terminate both ends of backplane
Arbitration group								
ARB5/-ARB0/	Open collector		60	0.1	0.9	220	330	Terminate both ends of backplane
BREQ/	Open collector		60	0.1	0.9	220	330	Terminate both ends of backplane
Central control group								
BCLK/	TTL	3	60	0.1	1.5	110	220	One point, furthest from driver
CCLK/	TTL	3	60	0.1	1.5	110	220	One point, furthest from driver
IDLACH/				0.1	1			See ADnn/ Lines
RESET/	TTL	3	18	0.1	1	330	470	Terminate both ends of backplane
PROT/	TTL	3	48	0.1	1	330	470	Terminate both ends of backplane
DCLOW/	TTL	3	48	0.1	1	330	470	Terminate both ends of backplane
RSTNC/	Open collector		60	0.1	0.9	220	330	Terminate both ends of backplane

Figure 5-12 MULTIBUS II electrical specifications.

are returned inactive, the arbitration process for a new master begins. This is the recovery phase and must last a minimum of three BCLK/ periods. As soon as a new master has gained ownership of the bus, it may resume transfer operations. This process is illustrated in Figure 5-11.

5.6 Electrical Specifications

The backplane is designed to be terminated so that ringing may be reduced to an acceptable level. A summary of the termination requirements for MULTIBUS II lines is shown in Figure 5-12, which also shows the drive requirements for each set of output lines and the type of output for each set of lines. Receiver specifications are also shown.

6

VMEbus

6.0 VMEbus Overview

In the realm of standard commercial busses, as of this date, the VMEbus is probably state of the art. It embodies most of the most powerful features available in bus architecture today, uses a wide data and address bus, and runs fast. It is being used in a number of new designs and most of the new CISC (complex instruction set computer) and RISC (reduced instruction set computer) processors are first being adapted to VMEbus boards. The VMEbus was originally conceived by Motorola's European microsystems group in Munich, West Germany, in the late 1970s and was refined and improved until it was adopted as an American national standard by ANSI and IEEE. The standard has been published as ANSI/IEEE Std 1014-1987. A copy of the standard is available through IEEE at 345 East 47th Street, New York, New York 10017. The mechanical specifications of the motherboard, backplanes, racks, and enclosures are based on IEC standard 297.

The original design was based on VERSAbus boards being built by Motorola at that time and was renamed VERSAbus-E. After prototype boards had been built and tested, Motorola, Mostek, and Signetics agreed to develop and support the new standard and a draft specification was written shortly afterward. The first revision "A" releases of the specifications were made available in October 1981 at the Systems 81 show in Munich. IEEE began work on the P1014 specification in 1983 with the final draft being adopted in June 1987.

Because of the complexity and diversity of VMEbus, it is difficult to cover all possible operations and implementation schemes that can be used on it. This chapter does not take the place of the P1014 specification and any designer should obtain a copy of the specification before proceeding. It is hoped that the information here will provide a basis for familiarization and emphasize the more common types of bus operations and implementations that occur on VMEbus.

6.1 Basic Features and Capabilities of VMEbus

The features supported by VMEbus include

1. Multiple masters with multiple arbitration schemes
2. Data transfers to 40 Mbytes/s
3. 32-bit data bus
4. 32-bit address bus
5. Multilevel interrupt capability
6. Dynamic bus sizing allowing 8-, 16-, and 32-bit devices
7. Several standard mechanical board sizes
8. Bus error detection and handling

VMEbus is broken into several subgroup busses with each containing a collection of lines or signals that perform a specific set of functions. There can be shared lines that perform multiple functions for different types of bus operations. For example, some address lines are used to transfer address information during data transfer cycles and also contain interrupt vector information during interrupt operations.

The major subgroups of lines that perform a specific set of functions are

1. *Data transfer bus:* Used to transfer data between masters and slaves.
2. *Priority interrupt bus:* Used to perform and handle interrupts.
3. *Utility bus:* Handles a variety of "clean up" functions such as power failure detection and handling, system clock, and initialization.
4. *Data transfer arbitration bus:* Used to control and determine bus ownership by a master. This bus arbitrates bus requests and handles bus master exchange operations.

6.2 VMEbus Mechanical Specifications

The complete mechanical specification of VMEbus includes rack dimensions, backplane data, board dimensions, connection details, and other information. The two most common types of boards encountered are the

6.2 VMEbus Mechanical Specifications

single-height board (or 3U board) and the double-height board (or 6U board). The single high board uses only a 24-bit address bus and 16-bit data bus. It is implemented using a single 96-pin connector on a board whose dimensions are 6.30 by 3.94 in. The outline of the single-height board is shown in Figure 6-1 and the bus is implemented on the P1 connector. The double-height board uses a 32-bit address bus and 32-bit data bus. It is implemented using two 96-pin connectors on a board whose dimensions are 6.30 by 9.19 in. The outline of a double height board is shown in Figure 6-2 and the bus is implemented on the P1 and P2 connector. All boards have a thickness of 0.062 in.

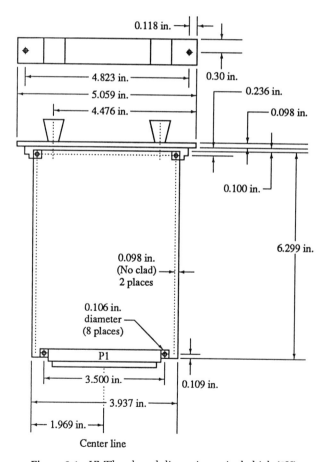

Figure 6-1 VMEbus board dimensions, single high (3U).

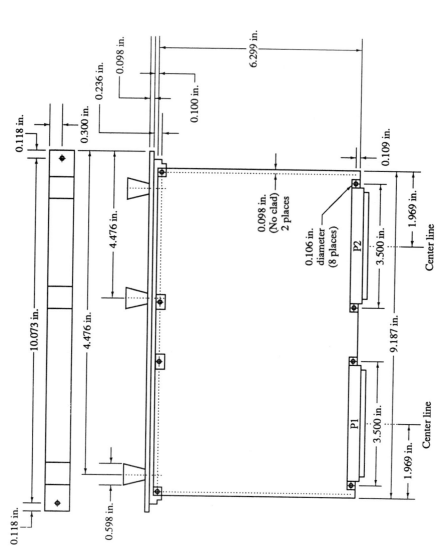

Figure 6-2 VMEbus board dimensions, double high (6U).

6.3 VMEbus Mechanical Specifications

6.2.1 Pin Assignments

The double-height board has a total of 192 pins available for use. Not all pins are used even with the full bus standard. The unused pins are left for individual designers to use as desired. These are user-defined (UD) pins. Both the P1 and P2 connectors are identical and consist of 3 rows of 32 pins. The rows are labeled A, B, and C rows, while the individual pins are labeled 1 through 32. A given pin is identified by an alphanumeric number indicating the row and pin, such as B21. The complete VME pin assignment for both P1 and P2 is shown in Table 6-1.

TABLE 6-1
VME Pin Assignment

Pin	P1 row A	P1 row B	P1 row C	P2 row A	P2 row B	P2 row C
1	D00	BBSY*	D08	User defined	+5 VDC	User defined
2	D01	BCLR*	D09	User defined	Ground	User defined
3	D02	ACFAIL*	D10	User defined	Reserved	User defined
4	D03	BG0IN*	D11	User defined	A24	User defined
5	D04	BG0OUT*	D12	User defined	A25	User defined
6	D05	BG1IN*	D13	User defined	A26	User defined
7	D06	BG1OUT*	D14	User defined	A27	User defined
8	D07	BG2IN*	D15	User defined	A28	User defined
9	Ground	BG2OUT*	Ground	User defined	A29	User defined
10	SYSCLOCK	BG3IN*	SYSFAIL*	User defined	A30	User defined
11	Ground	BG3OUT*	BERR*	User defined	A31	User defined
12	DS1*	BR0*	SYSRESET*	User defined	Ground	User defined
13	DS0*	BR1*	LWORD*	User defined	+5 VDC	User defined
14	WRITE*	BR2*	AM5	User defined	D16	User defined
15	Ground	BR3*	A23	User defined	D17	User defined
16	DTACK*	AM0	A22	User defined	D18	User defined
17	Ground	AM1	A21	User defined	D19	User defined
18	AS*	AM2	A20	User defined	D20	User defined
19	Ground	AM3	A19	User defined	D21	User defined
20	IACK*	Ground	A18	User defined	D22	User defined
21	IACKIN*	SERCLK	A17	User defined	D23	User defined
22	IACKOUT*	SERDAT	A16	User defined	Ground	User defined
23	AM4	Groune	A15	User defined	D24	User defined
24	A07	IRQ7*	A14	User defined	D25	User defined
25	A06	IRQ6*	A13	User defined	D26	User defined
26	A05	IRQ5*	A12	User defined	D27	User defined
27	A04	IRQ4*	A11	User defined	D28	User defined
28	A03	IRQ3*	A10	User defined	D29	User defined
29	A02	IRQ2*	A09	User defined	D30	User defined
30	A01	IRQ1*	A08	User defined	D31	User defined
31	-12 VDC	+5 V (standby)	+12 VDC	User defined	Ground	User defined
32	+5 VDC	+5 VDC	+5 VDC	User defined	+5 VDC	User defined

6.3 VMEbus Cycles

VMEbus also supports the following types of cycles:

Read Cycle An operation using the data transfer bus which causes the transfer of 1, 2, or 4 bytes of data from a slave module to a master. The master broadcasts address and address modifier information to all slaves. Each slave determines if it is being selected during the read cycle. The selected slave places data on the data transfer bus which is captured by the master initiating the transfer.

Write Cycle An operation using the data transfer bus which causes the transfer of 1, 2, or 4 bytes of data from a master to a slave. The master broadcasts address, data, and address modifier information to all slaves. If a slave is selected, it takes the data and loads them into the selected device.

Block Read Cycle An operation using the data transfer bus which permits the transfer of 1 to 256 bytes of data from a slave to a master. The transfer can be done by transferring any of 1, 2, or 4 bytes of data per transfer. The master retains control of the data transfer bus throughout the block transfer. The master broadcasts only the initial address and address modifier at the beginning of the block read. The slave then increments to the next sequential address on each succeeding data transfer operation. The use of a single address broadcast by the master to transfer the entire block of data makes this type of transfer faster and different from a sequence of read cycles.

Block Write Cycle An operation using the data transfer bus which permits the transfer of 1 to 256 bytes of data from a master to a slave. The transfer can be done by transferring any of 1, 2, or 4 bytes of data per transfer. The master retains control of the data transfer bus throughout the block transfer. The master broadcasts only the initial address and address modifier at the beginning of the block read. The slave then increments to the next sequential address on each succeeding data transfer operation. The use of a single address broadcast by the master to transfer the entire block of data makes this type of transfer faster and different from a sequence of read cycles. The process is identical to the block read with the exception of the direction of the data transfer.

Read-Modify-Write Cycle A bus cycle using the data transfer bus which permits a master to read from a given location, modify the data obtained, and write the data back to that location without allowing any other master to access the data being changed. This type of operation is particularly useful in multiple master systems in which software flags, mailboxes, or semaphores are used for interprocessor communication.

Interrupt Acknowledge Cycle A data transfer bus cycle that is generated by an interrupt handler. It is used to obtain information from the interrupting module that identifies the interrupter.

Address Only Cycle Used by a master to broadcast address-only information with no corresponding data transfer. There is no handshake between the master and addressed slave for this operation.

6.4 VME Functional Modules

VMEbus also breaks down operations as being performed by a "functional module." Each functional module is a set of circuitry residing on one board that performs a specific task. The functional modules supported by VMEbus are

Master A module that initiates data transfer bus cycles so that data may be moved between a slave and master.

Slave A module that responds to data transfer bus cycles initiated by a master module for the purpose of moving data between the master and slave.

Location Monitor A module that determines when an access to an address or range of addresses it decodes has been performed by a master. When access to one of these addresses occurs, an on-board signal is generated, notifying the board of the master's intended access.

Bus Timer A module that keeps track of the time required for a given data transfer to occur. If the time becomes excessive, a bus error condition is generated. This would occur if a master accesses a nonexistent slave.

Interrupter This module generates an interrupt using the priority interrupt bus. When an interrupt acknowledge is generated by an interrupt handler module, the interrupter places status/ID information on the bus in accordance with the interrupt handling protocol.

IACK Daisy-Chain Driver This module drives the interrupt acknowledge daisy chain after an interrupt handler acknowledges an interrupt request. This function ensures that only one interrupter will respond to the interrupt acknowledge cycle when several interrupters may be generating interrupt requests.

Requester This module is a part of the multiple master implementation scheme. Both masters and interrupt handlers use this module to request control of the VMEbus when needed.

Arbiter This module determines or arbitrates the various VMEbus requests, determining which bus master will gain control of the bus.

System Clock Driver A module that drives a 16 MHz utility clock signal on the bus.

Power Monitor This module determines the status or "health" of the main supply voltages to VMEbus. When out-of-specification power conditions occur on the bus, a signal is generated so that appropriate action can be taken.

A block diagram showing the bus modules and the busses each uses is shown in Figure 6-3.

6.5 Data Transfer Bus

All transfers performed on the VME data transfer bus are asynchronous. The data transfer bus consists of 34 address lines, 6 address modifier lines, 32 data lines, and 6 control lines. The lines are

A01–A31 Thirty-one address lines that form the address of the slave being selected by the current bus master initiating a data transfer cycle. The total address space of VMEbus is 2E32 bytes. It is important to note that there are only 31 address lines so that the address bus addresses one-word boundaries only. VME still supports byte transfers, however. The byte addresses are formed using A01–A31 lines in conjunction with DS0*, DS1*, and LWORD*. The address lines are active high and are driven by the current bus master.

AM0–AM5 The six tristate address modifier lines are used to qualify the address being generated by the bus master. These lines indicate the type of transfer that is going to be performed by the current bus master. There are potentially 64 address modifier codes. Of the 64 codes, 14 are predefined by VMEbus specification, 34 are reserved for future use by the bus specification, and 16 can be defined by any user.

By appropriate use of the address modifier codes, the bus can support bus masters who address in 16-bit, 24-bit, and 32-bit address space. These address spaces are referred to as short, standard, and extended address space, respectively. The 68000 series of processors have two operating modes called the supervisory and nonprivileged modes. The address modifier code indicates the type of transfer mode in which the processor is currently operating. Additionally, the address modifier code indicates whether the machine is performing an op-code or instruction fetch cycle or a data fetch cycle and whether the operation is a block transfer or standard transfer. A complete listing of address modifier codes is shown in Table 6-2. The address modifier lines are active high and are driven by the current bus master.

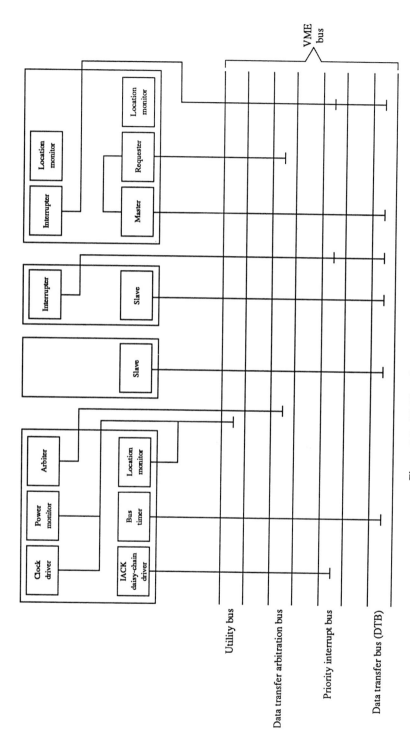

Figure 6-3 VMEbus functional modules.

TABLE 6-2
VMEbus Address Modifier Codes[a]

Address modifier code (hex)	Transfer type	Address modifier code (hex)	Transfer type
00	Reserved	20	Reserved
01	Reserved	21	Reserved
02	Reserved	22	Reserved
03	Reserved	23	Reserved
04	Reserved	24	Reserved
05	Reserved	25	Reserved
06	Reserved	26	Reserved
07	Reserved	27	Reserved
08	Reserved	28	Reserved
09	Extended nonprivileged data access	29	Short nonprivileged access
0A	Extended nonprivileged program access	2A	Reserved
0B	Extended nonprivileged block transfer	2B	Reserved
0C	Reserved	2C	Reserved
0D	Extended supervisory data access	2D	Short supervisory access
0E	Extended supervisory program access	2E	Reserved
0F	Extended supervisory block transfer	2F	Reserved
10	User defined	30	Reserved
11	User defined	31	Reserved
12	User defined	32	Reserved
13	User defined	33	Reserved
14	User defined	34	Reserved
15	User defined	35	Reserved
16	User defined	36	Reserved
17	User defined	37	Reserved
18	User defined	38	Reserved
19	User defined	39	Standard nonprivileged data access
1A	User defined	3A	Standard nonprivileged program access
1B	User defined	3B	Standard nonprivileged block transfer
1C	User defined	3C	Reserved
1D	User defined	3D	Standard supervisory data access
1E	User defined	3E	Standard supervisory data access
1F	User defined	3F	Standard supervisory block transfer

DS0* & DS1* The active low, tristate data strobe lines perform two functions. The level of the lines in conjunction with A01 and LWORD* are used to form the address of a slave being accessed by the master and the transitions are used to indicate the initiation or termination of the data transfer bus cycle. A summary of the levels for DS1*, DS0*, LWORD*,

6.5 Data Transfer Bus

and A01 for the common types of bus transfer operations are shown in Table 6-3. The data strobe lines are active low and are driven by the current bus master.

The high-to-low transition of the data strobe line(s) indicates that the master has placed valid data on the data lines during a write operation. The low-to-high transition of the line(s) indicates that the slave may release the data lines during a read operation and the master has taken the data.

LWORD* This active low, tristate line is used in conjunction with DS0* and DS1* to indicate the width of the data bus during a data transfer bus cycle. A low on this line means that the master is performing a long-word or quad byte transfer. An inactive high state on this line indicates that the master is performing a word or byte transfer operation.

D00–D31 These 32 bidirectional tristate data lines can be driven by either the bus master or a slave. During read operations, the lines are driven by the slave and during write operations, the lines are driven by masters. These lines are active high and contain the data to be moved between the master and slave. Not all data lines may be used on every type of transfer operation. Depending on the type of transfer, the 8, 16, or 32 data line may be used for byte, word, or long-word (quad byte) transfers. The VME data bus is dynamically configured on each transfer for the size of the transfer being performed.

AS* This line is an active low, tristate line driven by the bus master. The high-to-low transition of this line indicates valid address information on A01 through A31.

DTACK* An active low line driven by the slave responding to a master bus transfer cycle. Normally inactive high, this line is driven low when the slave has taken data from the data bus in the case of write operations or when the slave has placed valid data on the data lines in the case of read operations.

WRITE* This active low, tristate line is driven by the bus master and indicates the direction of a data transfer. The line is driven low to indicate a write operation or a transfer of data from the master to a slave. It is driven high to indicate a read operation or a transfer of data from the slave to the master.

BERR* The tristate, active low bus error line is used to indicate a type of transfer that cannot be performed by the slave or some other error condition that may occur on the bus, such as a time-out condition. The BERR* line is driven low by either a slave or bus timer. The BERR* line is a convenience, not required for all implementations. It is convenient for software debugging and the detection of hardware failures.

TABLE 6-3
Common Types of Bus Transfer Operations

Type of cycle	DS1*	DS0*	A01	LWORD*
ADDRESS-ONLY	high	high	<------Note1------>	
Single even byte transfers				
BYTE(0) READ or WRITE	low	high	low	high
BYTE(2) READ or WRITE	low	high	high	high
Single odd byte transfers				
BYTE(1) READ or WRITE	high	low	low	high
BYTE(3) READ or WRITE	high	low	high	high
Double byte transfers				
BYTE(0-1) READ or WRITE	low	low	low	high
BYTE(2-3) READ or WRITE	low	low	high	high
Quad byte transfers				
BYTE(0-3) READ or WRITE	low	low	low	low
Single byte block transfers				
SINGLE BYTE BLOCK READ or WRITE	<----------Note 2---------->			high
Double byte block transfers				
DOUBLE BYTE BLOCK READ or WRITE	low	low	Note 3	high
Quad byte block transfers				
QUAD BYTE BLOCK READ or WRITE	low	low	low	low
Single byte RMW transfers				
BYTE(0) READ-MODIFY-WRITE	low	high	low	high
BYTE(1) READ-MODIFY-WRITE	high	low	low	high
BYTE(2) READ-MODIFY-WRITE	low	high	high	high
BYTE(3) READ-MODIFY-WRITE	high	low	high	high
Double byte RMW transfers				
BYTE(0-1) READ-MODIFY-WRITE	low	low	low	high
BYTE(2-3) READ-MODIFY-WRITE	low	low	high	high
Quad byte RMW transfers				
BYTE(0-3) READ-MODIFY-WRITE	low	low	low	low
Unaligned transfers				
BYTE(0-2) READ or WRITE	low	high	low	low
BYTE(1-3) READ or WRITE	high	low	low	low
BYTE(1-2) READ or WRITE	low	low	high	low

1. During ADDRESS-ONLY cycles, both data strobes are maintained, but the A01 and LWORD* lines might be either high or low.
2. During single-byte block transfers, the two data strobes are alternately driven low. Either data strobe might be driven low on the first transfer. If the first accessed byte location is BYTE(0) or BYTE (2), then DS1* is driven low first. If the first accessed byte location is BYTE(1) or BYTE(3), then DS0* is driven low first. A01 is valid only on the first data transfer (i.e., until the SLAVE drives DTACK* or BERR* low the first time) and might be either high or low, depending on which byte the single-byte block transfer begins with. If the first byte location is BYTE(0) or BYTE(1), then A01 is low. If the first byte location is BYTE (2) or BYTE (3), then A01 is high.
3. During a double-byte block transfer, the two data strobes are both driven low on each data transfer. A01 is valid only on the first data transfer (i.e., until the SLAVE drives DTACK* or BERR* low the first time) and might be either high or low, depending on what double-byte group the double-byte block transfer begins with. If the first double-byte group is BYTE (0–1), then A01 is low. If the first double-byte group is BYTE (2–3), then A01 is high.

6.5.1 Data Transfer Bus Operations

There are a number of data transfer operations that may be performed on the DTB. They are

1. Single byte read or write
2. Double byte (word) read or write
3. Quad byte (long-word) read or write
4. Address-only broadcast
5. Read-Modify-Write
6. Block read or write
7. Unaligned word or long-word transfers

Unaligned Transfers

In addition to the above DTB operations, transfers may be unaligned. Random access memory (RAM) is organized and addressed on a byte basis, starting at address 0, 1, 2, 3, 4, 5, 6, 7 . . . , etc. Each address may be accessed by a master for the purpose of reading or writing a byte of data into RAM. When 16 bits or a word of data is stored in memory it is "normally" stored at a word boundary. Word boundaries are at even byte addresses, so words would normally be stored at 0, 2, 4, 6, 8 . . . , etc. In like manner, when 32 bits or a long word of data is stored in memory, it is "normally" stored at long-word boundaries. Long-word boundaries are at 0, 4, 8, 12 . . . , etc. VMEbus allows for the storage of words at nonword boundaries and long words at nonlong-word boundaries by use of multiple byte transfer cycles. The storage of word or long-word data at byte boundaries that are not word or long-word boundaries are called unaligned transfers.

The acceptable sequences by which a long word may be stored at boundaries that are not long-word aligned are

I. Long-word transfer starting at addresses 1, 5, 9, etc.
   A. Using a 16-bit data bus
      1. Byte on D00–D07 (store low 8 bits)
      2. Word on D00–D15 (store middle 16 bits)
      3. Byte on D08–D15 (store upper 8 bits)
   B. Using a 32-bit data bus
      1. Triple byte on D00–D23 (store low 24 bits)
      2. Byte on D08–D15 (store upper 8 bits)
II. Long-word transfer starting at addresses 2, 6, 10, etc.
   A. 16- or 32-bit data bus
      1. Word on D00–D15 (store low 16 bits)
      2. Word on D00–D15 (store upper 16 bits)

III. Long-word transfer starting at addresses 3, 7, 11, etc.
   A. Using a 16-bit data bus
      1. Byte on D00–D07 (store low 8 bits)
      2. Word on D00–D15 (store middle 16 bits)
      3. Byte on D08–D15 (store upper 8 bits)
   B. Using a 32-bit data bus
      1. Byte on D00–D07 (store low 8 bits)
      2. Triple byte on D08–D31 (store upper 24 bits)

The acceptable sequences by which a word may be stored at boundaries that are not word aligned are

I. Word transfer to addresses 1, 5, 9, etc.
   A. Using a 16-bit data bus
      1. Byte on D00–D07 (store lower 8 bits)
      2. Byte on D08–D15 (store upper 8 bits)
   B. Using a 32-bit data bus
      1. Word on D08–D23 (store full 16 bits)
II. Word transfer to address 3, 7, 11, etc.
   A. Using a 16- or 32-bit data bus
      1. Byte on D00–D07 (store lower 8 bits)
      2. Byte on D08–D15 (store upper 8 bits)

Note that all unaligned transfers may take more than one data transfer bus cycle and take more time to execute than aligned transfers. However, if unaligned transfers are not used, utilization of memory is not optimized and there will be some locations that are not used. Also, not all boards support unaligned transfers (UAT). A board vendor must explicitly state that the board supports UAT capability or it must be assumed that the board does not have the capability.

Address-Only Broadcast

An address-only broadcast does not transfer data. It is similar to a data transfer in every respect in that a valid address is driven onto the bus and the address strobe (AS*) is driven low. However, the data strobes (DS0*, DS1*) are never activated. The addressed slave does not respond by driving the data transfer acknowledge (DTACK*) low to complete the cycle. Address data is simply held on the address lines for a prescribed period. Address broadcasts are a speed-enhancement feature and are not supported by all masters. It allows slave boards to decode addresses concurrently with the bus master so that if the master accesses the addressed slave, the address decoding is already accomplished by the slave. This saves slave decoding time. As in the case of UAT, a board vendor

6.5 Data Transfer Bus

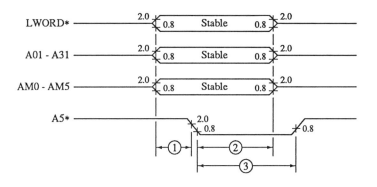

Parameter	Times (ns)		
	At master	At slave	At location monitor
1	35	10	10
2	40	30	30
3	40	30	30

Figure 6.4 VME address broadcast.

must explicitly state that a given master supports address-only (ADO) capability. An address-only broadcast timing diagram is shown in Figure 6-4.

Single-Byte Read and Write

All DTB transfers begin with the bus master driving the address and address modifier lines valid. At the same time that the address lines are driven valid, the interrupt acknowledge line (IACK*) must be inactive high for all data transfers and LWORD* must be inactive high for byte and word transfers. When the address is stable, the master drives AS* low, indicating to the slave that address decoding may begin. Additionally, WRITE* must be driven low for a transfer of data from the master to the slave (write) operation or high for a transfer of data from the slave to the master (read). The master must then specify the data width, which is a single-byte transfer in this case. A byte transfer is indicated by driving DS0* low and DS1* high if the byte transfer is to an even-byte address. DS1* is driven low and DS0* high if the byte transfer is to an odd-byte address. The slave being addressed selects the on-board device and fetches the data. Since a byte transfer cycle is specified by the master, the slave selects the byte value at the indicated address and presents the data on the D00–D07 lines in the case of a read or loads the data on D00–D07 into the selected device. The slave then drives the DTACK* line low, indicating that the data is stable or has been loaded. The master responds to DTACK* low by taking the data from D00–D07 (read operation only) or releasing D00–D07 (write operation). Also, the address lines, address modifier, LWORD*, and IACK* are released for further transfers. DS* and AS* are then returned to their inactive high state and the master may then address the next slave. The slave finishes the transfer cycle by releasing DTACK* to its inactive high state when it detects AS* and DS0* high. A timing diagram showing a single-byte transfer is shown in Figure 6-5.

Word (Double-Byte) Read and Write

A timing diagram showing a word transfer is shown in Figure 6-6. Word transfers are very similar to byte transfers with the exception of the DS0* and DS1* signals. During a word transfer, both DS0* and DS1* are driven low rather than just DS0* or DS1* being driven low. Also data are placed on D00 through D15 rather than just D00–D07 if the transfer is aligned.

An unaligned transfer requires either two-byte transfers or a single transfer cycle using D08 through D23, as discussed earlier in the section entitled Unaligned Transfers, (p. 93).

Long-Word (Quad-Byte) Read and Write

A timing diagram showing a long-word transfer is shown in Figure 6-7. Long-word transfers are nearly identical to byte transfers. All signals are handled the same with the exception of DS0*, DS1*, LWORD*, and D00 through D31. For a long-word transfer, all three of these lines are driven low by the bus master and, for aligned transfers, the 32-bit data is placed on D00 through D31.

In the case of unaligned transfers, the transfer may require several transfer cycles using both word and byte transfers, as discussed earlier in the section entitled Single-Byte Read and Write (p. 95).

Block Read and Write

Block data transfers allow for the movement of up to 256 bytes of data into consecutive memory locations. The timing diagram of a single-byte block write is shown in Figure 6-8. In block operations, only the starting address of the data transfer is broadcast at the beginning of the block transfer. A series of data strobes on DS1* and DS0* are generated by the

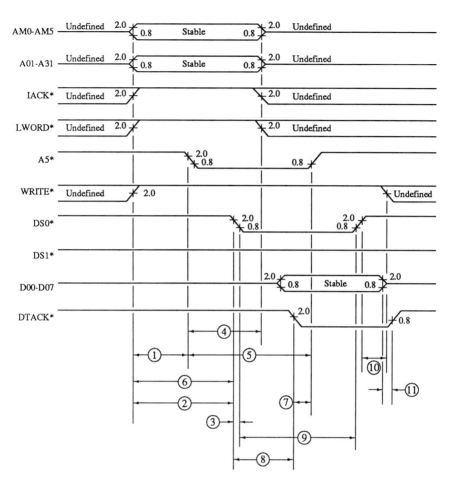

Figure 6-5 VMEbus data transfer bus single-byte read.

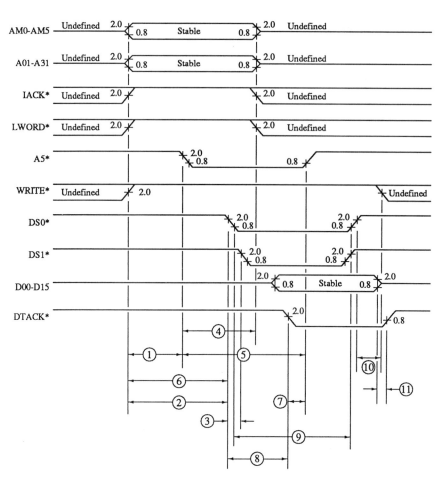

Figure 6-6 VMEbus data transfer bus word read.

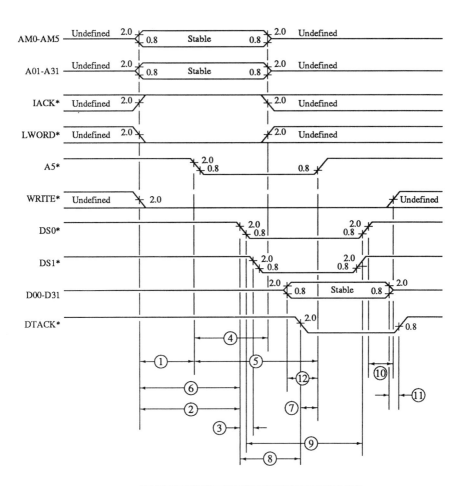

Figure 6-7 VMEbus data transfer bus long-word write.

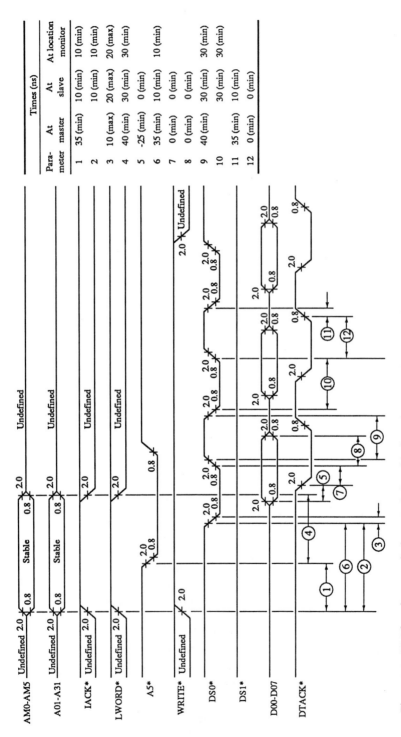

Figure 6-8 VMEbus data transfer bus, single-byte block read. (Note: 3 bytes transferred in this illustration; process extends up to 256 bytes.)

master. For write operations, data are also placed on the data bus by the master in a manner similar to a single-byte write. The slave takes the data, generating a DTACK* signal to complete each transfer. For read operations, data is placed on the data bus by the addressed slave following each data strobe. The DTACK* signal is generated by the slave to indicate to the master that stable data is on the bus. The use of DS0*, DS1*, and LWORD* to specify the size of the transfer and byte address is identical to single DTB transfer cycles.

Read-Modify-Write Operations

A read-modify-write (RMW) operation is simply a read operation followed immediately by a write operation to the same address. There is only one address broadcast at the beginning of the cycle. The timing diagram for a word RMW cycle is shown in Figure 6-9. The principal purpose of RMW is to facilitate the use of flags, mailboxes, or semaphores in multitasking systems. Errors can occur if one task is in the process of changing a semaphore to another task and a context switch occurs. An RMW cycle prevents task switching while shared resources are being modified.

The process begins with an address broadcast cycle in which the master generates valid address data followed by an active low on the AS* line. For the first part of the two-part process, the data are read from the slave so the WRITE* line is inactive high. The DS1*, DS0*, and LWORD* lines are used in a manner identical to normal byte, word, and long-word reads and writes. The responding slave places addressed data on the data lines and generates a low on DTACK*. The master receiving DTACK* low drives the DS1* and DS0* lines high, which in turn signals the slave to drive DTACK* high, completing the "read" part of the RMW cycle. The data is modified by the master and is now ready to write back to the same memory location. The master drives the same DS1*, DS0*, and LWORD* signals as for the "read" part of the RMW cycle. The slave responds by loading the data into the addressed device and driving DTACK* low. The master receives DTACK* low and releases DS1* and DS0* to their inactive high state. The slave receives DS1* and DS0* high and releases DTACK* high, completing the RMW cycle.

6.6 Data Transfer Arbitration Bus

The data transfer arbitration bus (DTB arbitration bus) is used in the VME system in which there are more than one bus master. The practice of

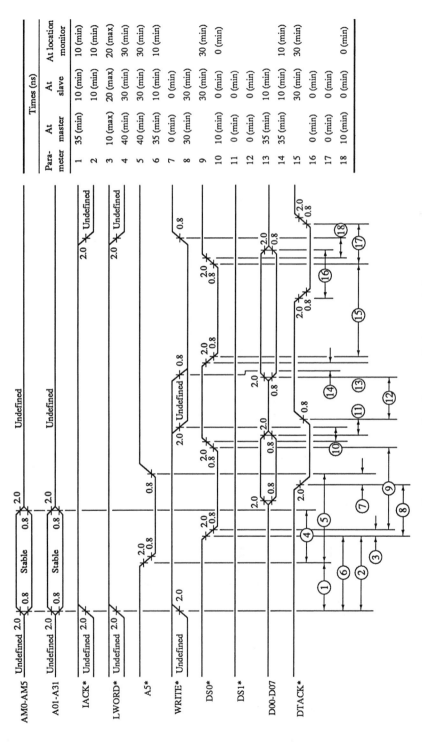

Figure 6-9 VMEbus data transfer bus, single-byte RMW.

6.6 Data Transfer Arbitration Bus

increasing a system's power by using distributed processing has become increasingly common in the past few years. The purpose of the arbitration bus is to prevent the simultaneous generation of DTB cycles by more than one bus master. It also implements one of a number of schedule algorithms so that efficiency in use of the bus may be improved. A master that may contend for use of the bus must have a requestor module that handles requests by the master. All multiple-master VME system busses have a bus arbiter. The bus arbiter implements the arbitration algorithm and controls which master has control of the bus.

6.6.1 DTB Arbitration Lines

There are four daisy-chained and six bussed lines that form the DTB arbitration bus. Each requestor must drive the daisy-chained bus grant in and bus grant out lines. The requestor closest to slot 1 of the VME rack has the highest serial priority. Each requestor's bus grant in line is connected to a higher priority master's bus grant out line and the bus grant out lines are connected to the lower priority bus grant in lines. There are four sets of these daisy-chain lines, allowing for the implementation of a two-dimensional prioritizing scheme. One dimension is a serial prioritizing method and the second is a line prioritizing method. This is shown in Figure 6-10. So the daisy-chained lines are

BG3IN*	BG3OUT*
BG2IN*	BG2OUT*
BG1IN*	BG1OUT*
BG0IN*	BG0OUT*

For each of these sets of lines to grant control of the bus, there must be a line to request control of the bus so there are four bus request lines called BR3*, BR2*, BR1*, and BR0*. Additionally, there are two other bussed lines used for general-purpose control called bus busy (BBSY*) and bus clear (BCLR*).

BBSY* is a single line driven by the master currently in control of the bus. When BBSY* is released inactive high, the arbiter may grant control of the bus to another requestor. BCLR* is driven low by the arbiter to indicate to the current master that a higher priority request for the bus is pending. The current master must then release the bus.

6.6.2 Arbitration Algorithms

There are a great number of arbitration algorithms that may be implemented within the general VME structure. However, there are three that are specifically discussed and supported in the VME standard. They are

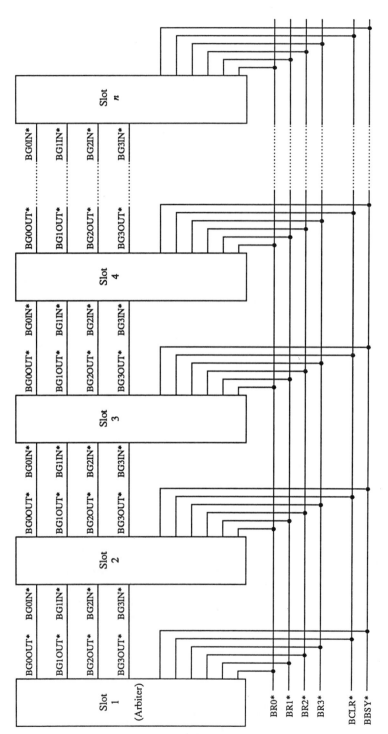

Figure 6-10 Data transfer bus arbitration lines.

6.6 Data Transfer Arbitration Bus

1. Single level
2. Prioritized
3. Round robin

There are four bus request lines that may be driven by any potential bus master.

In single-level (SGL) arbitration schemes, only the BR3*, BR3IN*, and BR3OUT* lines are used for arbitration. The two control lines BBSY* and BCLR* are used for all bus exchange operations. The arbiter, which is always in slot 1, must drive only these lines. Prioritization is achieved by the serial prioritization that occurs on the BR3IN*–BR3OUT* daisy chain. The method uses only one dimension, the serial arbitration method, of the potential two-dimensional bus arbitration scheme implemented by VMEbus. For example, if a single-level arbiter is installed in slot 1 and three single-level bus requestors installed in slots 5, 8, and 11 are all requesting use of the bus by generating a low on BR3*, the requester in slot 5 (closest to the arbiter) will be granted control of the bus. When the requester in slot 5 no longer needs the bus, the requestor in slot 8 will be granted control.

The prioritized (PRI) arbiter uses both dimensions of the bus arbitration methodology. The serial method is used on each of the BRn*, BGnIN*, and BGnOUT* lines. In addition to the serial arbitration on each set of daisy-chained lines, each set of lines is assigned a fixed priority. The highest priority is BR3*, BG3IN*, and BG3OUT*, descending in order to BR0*, BG0IN*, and BG0OUT* being the lowest priority. Assume there are four bus requestors installed in a VME rack which implements a prioritized arbitration scheme as follows:

Requester	BR*line driven	Installed slot
1	2	3
2	3	6
3	0	2
4	3	8

The order of priority for the bus requestors is

Priority	Requester
1 (highest)	2
2	4
3	1
4	3

First, the arbiter resolves which of the four BR* lines is being driven, with BR3* having the highest priority. The highest priority bus grant (BG3OUT*) line is driven active by the arbiter. Since there is more than one requester driving the BR3* line, the serial arbitration scheme assigns the highest priority to the requester closest to the arbiter or the requester in slot 6. In like manner the next highest priority is BR3*, slot 8, continuing down to BR0*, slot 2.

The round robin select (RRS) arbiter works slightly differently. The priority assigned to each BRn* line changes or rotates with each successive bus exchange operation. When the system starts, the order of priority from highest to lowest is BR3*, BR2*, BR1*, BR0*. Assuming there is a BR3* asserted upon initialization, the requestor having the highest serial priority driving BR3* will be granted control of the bus. After the requester driving BR3* relinquishes control of the bus, the arbiter rotates the priority so that it is now BR2*, BR1*, BR0*, with BR3* having the lowest priority. The requester having the highest serial priority driving BR2* will be the next to be granted control of the bus by the arbiter. After it relinquishes control of the bus the priority becomes BR1*, BR0*, BR3*, and BR2*. Now if there is no pending bus request on BR1*, the arbiter again rotates the priority to BR0*, BR3*, BR2*, and BR1*. This process continues until a new requester is granted control of the bus.

6.6.3 Bus Exchange

The process by which one master relinquishes control of the data transfer bus and another gains control is always a two-part operation. The first part is arbitration and the second part is exchange. VMEbus supports multiple arbitration algorithms, as discussed in the previous sections. At the end of the arbitration cycle, the arbiter has selected a new master that will be the next master in control of the bus. This is signaled to the new master by setting a BGnIN* active low. Since the new master desires control of the bus, it does not drive its corresponding BGnOUT* line low. The arbiter has selected the new master but the process of exchanging control of the bus is controlled by the new and old masters, not the arbiter.

In the following discussion, assume that an arbitration algorithm is implemented that gives the master driving BR2* a higher priority than the master driving BR1*. Assume also that both masters are contending for control of the bus at the same time. The process starts with both masters driving BR1* and BR2* active low, requesting control of the DTB. The arbiter detects both requests and in turn drives BG2IN* low, telling the master driving BR2* that it has been granted control of the bus. The BR2* master releases the BR2* line and drives the BBSY* line active low,

6.6 Data Transfer Arbitration Bus

indicating that it has control of the bus. After making sure AS* is inactive (high), master 2 (BR2*) begins performing data transfers over the DTB. When master 2 finishes using the bus, it releases BBSY* to its inactive high state. The arbiter detects that BBSY* is inactive and that there is still a request pending from master 1, which is still driving BR1* active. The arbiter then drives BG1IN* low, telling master 1 that it has control of the bus. Master 1 detects BG1IN* low and responds by driving BBSY* low, indicating to the arbiter that it has control of the bus. After setting BR1* to an inactive state and ensuring that AS* is inactive, master 1 begins performing data transfers over the DTB. When master 1 completes the data transfers, it releases BBSY*, indicating to the arbiter that the bus is free for use. The arbiter detects that BBSY* is inactive and waits for another bus request. The timing diagram for this process is shown in Figure 6-11.

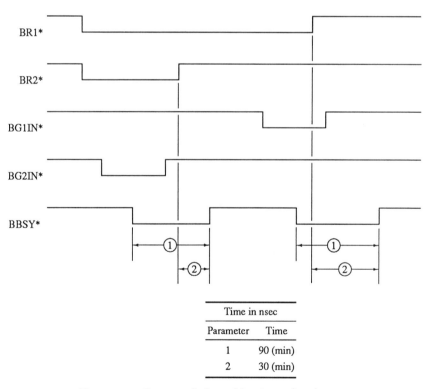

Time in nsec

Parameter	Time
1	90 (min)
2	30 (min)

Figure 6-11 Data transfer bus arbitration and exchange.

The bus exchange process is identical, regardless of the type of arbitration implemented. Requests at the same level are arbitrated by the serial or daisy-chain method. When the arbiter drives the BGnIN line active low, each successive master in the daisy chain not requesting use of the bus will drive the BGnOUT* at its output low also. When the master is requesting use of the bus receiving the BGnIN* line active, it holds the BGnOUT* line inactive and drives BBSY* active, indicating that it has control of the bus. It begins performing data transfers after releasing BRn* and receiving AS* inactive.

6.7 Priority Interrupt Bus

The priority interrupt bus provides a means by which an interrupter can generate interrupts and direct the servicing of those interrupts. There are two general methods by which interrupts are serviced.

There are single handler and distributed handler systems. In single handler systems, all interrupts are received and processed by a single processor or interrupt handler. Interrupts are prioritized so that a system design has latitude in the selection of interrupt types and the assignment of interrupt priority. Single handler interrupts are well suited to systems that require regular scheduling of events and continuous, uninterrupted execution by dedicated coprocessors. Distributed systems have more than one interrupt handler. Interrupt handlers may be implemented on several processor boards with each handling only certain bus interrupts. This type of system is useful in systems having independent processing functions being performed by multiple processors. It provides for a communication path when the independent processors must pass information between themselves.

Like many functions performed on VMEbus, the interrupt and interrupt handling process makes use of lines on the data transfer bus and the DTB arbitration bus. The priority interrupt bus in itself consists of nine lines. Eight of the lines are bussed and one is daisy-chained. The lines are

1. IRQ1*   Bussed
2. IRQ2*   Bussed
3. IRQ3*   Bussed
4. IRQ4*   Bussed
5. IRQ5*   Bussed
6. IRQ6*   Bussed
7. IRQ7 *   Bussed
8. IACK*   Bussed
9. IACKIN*/IACKOUT* Daisy-chained

6.7 Priority Interrupt Bus

IRQ1* through IRQ7* are the bus interrupt request lines driven by the interrupter. They are prioritized, with IRQ7* having the highest priority. The IACK* line is driven by interrupt handlers and is an input to the IACK daisy-chain driver module. This module is always located in VMEbus slot 1. The IACK* line is used to signal that an interrupt handler is responding to an interrupt request. The IACK daisy-chain driver in turn drives the IACKIN*/IACKOUT* daisy chain. The IACKIN*/IACKOUT* daisy chain ensures that only one interrupt handler responds to an interrupt request.

6.7.1 IACKIN*/IACKOUT* Daisy Chain

The IACKOUT* is driven by each higher priority interrupter and is received on the IACKIN* lower priority interrupters. If a slot in the VME rack is empty between an interrupter and the IACK daisy-chain driver in slot 1, the IACKIN*/IACKOUT* daisy-chain pins must be jumped on the backplane to allow the signal to ripple down the bus. The interrupter generating an interrupt that has the highest priority will pass the IACKOUT* daisy-chain signal down to the lower priority interrupter. Interrupt priority is determined by position in the bus, with the interrupter closest to the IACK daisy chain driver in slot 1 having the highest priority. A block diagram of this process is shown in Figure 6-12.

6.7.2 Interrupt Process

The interrupt process begins with the generation of an interrupt request generated by an interrupter. The interrupt is prioritized by the IRQ line, with IRQ7* being the highest priority. In single handler systems, all IRQ lines being used are monitored by a single interrupt handler. In distributed handler systems, only a set of those lines is monitored by each interrupt handler. If an interrupt request is directed to an interrupt handler that is not currently in control of the bus, the interrupt handler must proceed with a bus arbitration and exchange process and gain control of the bus before servicing the interrupt. See Sections 6.6 through 6.6.3 for an explanation of the process by which bus arbitration occurs. Once the interrupt handler has control of the bus, the handler generates an interrupt acknowledge receiving STATUS/ID information from the interrupter. It uses the STATUS/ID information to initiate appropriate interrupt servicing. Interrupt handlers can be any of three types. A D08 handler handles STATUS/ID using only byte transfers, D16 uses 16-bit transfers, and D32 can use 32-bit transfers from the interrupter. Note that the interrupt handler does not write to the interrupter during the interrupt acknowledge cycle and does not drive the WRITE* line. The servicing of an interrupt is implementation-dependent. There are no requirements

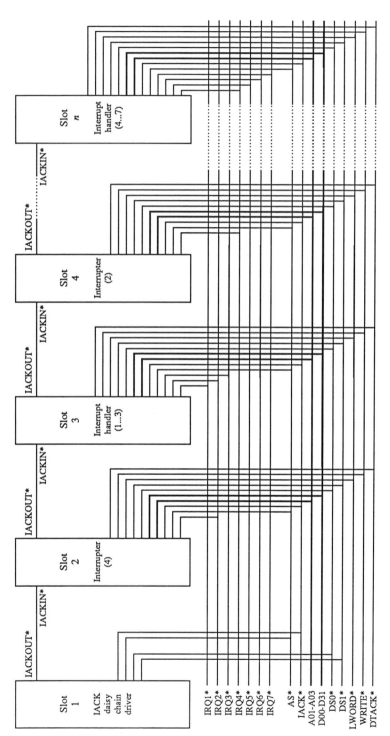

Figure 6-12 VME priority interrupt bus.

that VMEbus places on the interrupt service routine and that routine may or may not require use of the bus.

Interrupt Handler Operation

The interrupt handler acknowledges the bus interrupt by generating an interrupt acknowledge in which IACK* is driven active low. It places the encoded value of the interrupt it is acknowledging on address lines A01 through A03. For example, if the interrupt handler is acknowledging an interrupt on IRQ5*, A03 through A01 would be driven with the value 101 (binary) or 5. Additionally, the interrupt handler need not drive the address modifier lines or data lines, unlike a normal bus master.

Interrupter Operation

The interrupting module generates an interrupt by driving the IRQn* line active low. When the interrupt acknowledge daisy chain goes active low and A03 through A01 contain the same code as the interrupt line being driven, it responds by placing STATUS/ID on the bus data lines. The IACK* line functions in a manner similar to AS* during data transfers. The falling edge of IACK* indicates valid data on A03 through A01. Note also that slave modules must monitor IACK* to ensure that they do not respond to an interrupt acknowledge cycle. IACK* must be high (inactive) during a DTB cycle and low (active) during an interrupt acknowledge cycle. The interrupt handler must also drive DS1*, DS0*, and LWORD* to indicate to the interrupter the width of the STATUS/ID transfer. The interpretation of these three lines is identical to the way they are used for data transfer bus operations. If an interrupt handler requests a STATUS/ID transfer of a size greater than the size the interrupter is capable of generating, the interrupter still responds with the STATUS/ID it is capable of generating. All data lines left undriven by the interrupter will be interpreted as high by the interrupt handler due to the presence of bus terminating resistors that pull the bus data lines high in the absence of drive.

VMEbus supports two methodologies by which the interrupter releases the IRQn* line back to the inactive state. The first method is called ROAK or release on acknowledge. After the interrupter transfers STATUS/ID information to the interrupt handler, the interrupter will release the driven IRQn* line. The second is called RORA or release of register access. When the interrupt handler enters the interrupt service routine, instructions access key registers within the interrupter with either a read or write operation. This register access causes the interrupter to release the driven IRQn* line.

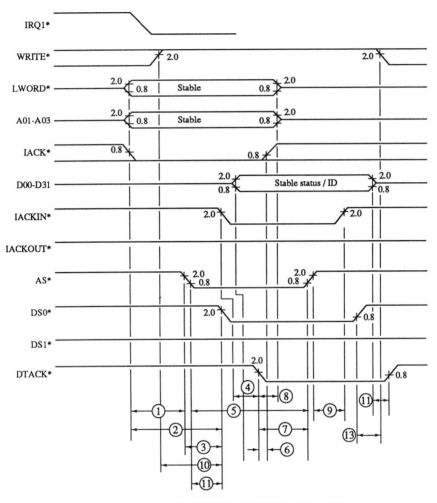

Figure 6-13 VMEbus interrupt and interrupt acknowledge cycle, single byte.

IACK Daisy-Chain Driver Operation

The IACK daisy-chain driver generates a falling edge on the IACKIN*/IACKOUT* daisy chain when the interrupt handler initiates an interrupt acknowledge cycle. It monitors the IACK*, AS*, DS1*, and DS0* lines. When IACK* is driven low, AS* is inactive (high), and DS1*, DS0* are both driven by the interrupt handler, the daisy-chain driver generates an active edge on IACKOUT* to the highest priority master in the daisy chain.

Timing diagrams showing a byte, word, or long-word interrupt and interrupt acknowledge cycle assuming a single handler system are shown in Figure 6-13.

7

NuBus

7.0 NuBus Overview

The NuBus standard has been adopted as a national standard and is published by IEEE as ANSI/IEEE Std 1196-1987. Also, like several other busses, the mechanical configuration of NuBus is consistent with the requirements of ANSI/IEEE 1101-1987. This standard defines the mechanical specifications for microcomputers. The intent of NuBus is to implement a 32-bit high-speed interface that uses a very simple and direct design methodology. It is not oriented nor was the design driven by any particular CPU. The bus concept was originated at MIT in 1979 by Professors Steve Ward and Chris Terman. Their early work resulted in the support of Western Digital Corporation. The combined effort of MIT and Western Digital Corporation resulted in the core of the current specification, which was developed in 1981. In 1983, the work was moved to Texas Instruments Corporation for further development with the hope of merging the NuBus effort with Future Bus (P896). The prospect of merging the two busses became unlikely and a separate NuBus group, which produced the Std 1196 specification, was formed in 1984.

NuBus is optimized for 32-bit transfers. It runs at a transfer rate of 10 MHz and is a synchronous bus. It supports multiple masters and a methodology for arbitrating multiple masters that is "fair." The principal features of the bus include the following:

1. The only bus transfers supported are read, write, block read, and block write.
2. 32-bit address space.
3. Geographic slot addressing.

4. Simple board configuration.
5. Minimal pin count—only 51 signal lines.
6. CPU-independent.
7. Multiple masters.

The NuBus concept uses the same master and slave module concept as that used by most busses. A master originates but transfer operations and a slave responds to bus operations.

All signal lines on NuBus are active low signals. A high voltage state is the unasserted state or logical false, while a low voltage is the asserted state or logical true condition.

7.1 NuBus Mechanical Specifications

Two board sizes are specified under the NuBus specification. There is a "triple-height" board and a "PC"-style board.

7.1.1 NuBus Triple-Height Board Specifications

The triple-height form factor is shown in Figure 7-1. There are three connectors specified in the triple-height form factor; however, all of the NuBus signals are carried on the P1 connector. The connectors are C096-style connectors like those used on VMEbus. The component height above the mounting surface of the board is to be less than 0.55 in. Component or lead length from the bottom of the board is specified at less than 0.10 in. The board thickness is nominally specified as 0.063 in.; however, thicker boards may be used if the boards are 0.063 in. thick and 0.10 in. on the edge of the board where the board is supported by the card cage guides. This requires milling the edge of the board during fabrication, which adds to the board fabrication cost. If the boards are made thicker, the added thickness must be subtracted from the bottom component and lead clearance specification. Parts may not be placed within 0.150 in. from the top or bottom edge of the board to prevent interference with card guides.

Indicators may be placed on the board but the location is controlled by the specification. They must be located in the lower left corner, as shown in Figure 7-1.

The specification also requires that injection/ejection tabs be installed on all boards. The card cage is provided with a specially designed surface that engages the tabs and allows for easy insertion and removal of boards. All boards are spaced on 0.80 in. centers within the card cage.

7.1 NuBus Mechanical Specifications

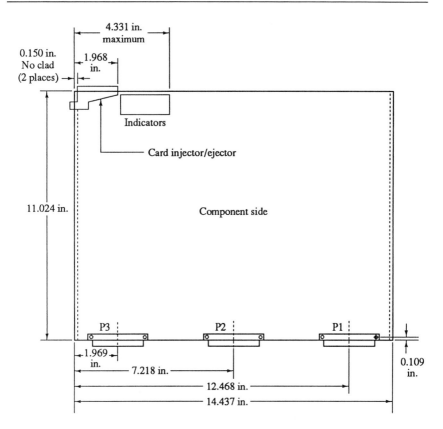

Figure 7-1 NuBus triple-height form factor board outline.

Note also that the specification does not permit the connection of external cabling from the front edge of the board. All cabling must be connected to the P2 or P3 user-defined connectors located on the back (P1) side of the board.

The P2 and P3 connectors are largely undefined by the specification. Eighteen pins on the P2 connector are defined for bus use. The specification also requires that only TTL I/O signals be applied to the A and C rows of the connector, while unused pins on the B row may be used for non-TTL signals.

7.1.2 PC Board Specifications

The PC form factor NuBus board outline is shown in Figure 7-2. There is a significant difference in that a single connector is on the NuBus

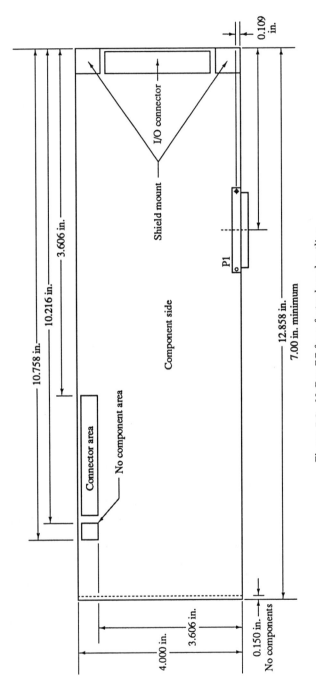

Figure 7-2 NuBus PC form factor board outline.

connector side of the board and one I/O connector is defined for use on the opposite side. An expansion shield similar to the type used on PC/AT-style machines is specified for one end of the board. The basic size of the PC form factor NuBus board is 12.86 (maximum) by 4.00 in. The 12.86 in. dimension can be made smaller for any given implementation but that dimension may not be less than 7.00 in. The style connector (C096) is the same as that of the triple-height board and the pin assignment is the same. The specification does not control the number and size of the I/O connectors. The auxilliary connector area shown on the outline drawing is for guidance only. The PC form factor board component height limitation is 0.60 in. and lead and component height on the other side.

The board thickness is again defined as 0.063 in. No components may be placed within 0.15 in. of the end of the board engaging the card guide to prevent interference with the guide. All card cages using the PC form factor place boards on 0.90 in. centers.

7.2 Bus Lines

NuBus is subdivided into the following groups of signals, with each group of lines being used to perform a given function.

7.2.1 Power

There are four voltages defined by the bus specification. They are

1. +5 VDC
2. +12 VDC
3. −12 VDC
4. −5.2 VDC

The maximum current load is not controlled by specification. The −5.2 VDC power is optional. If it is not implemented in a given bus, the lines should not be used for any other function. The tolerance on all power supplies is ± 3.0%. The line and load regulation must be better than 0.3%. The peak-to-peak ripple on the +5 and −5.2 VDC supply lines should not be greater than 50 MV. The peak-to-peak ripple on the +12 and −12 VDC power supply lines should not be greater than 75 MV.

7.2.2 Utility Signals

CLK/ A 10-MHz, 25% duty cycle signal used to synchronize all data transfer and bus arbitration operations. All bus signals are in transition on the rising edge of the clock and sampled on the falling edge.

RESET/ This open collector, active low signal returns all devices on the bus to a power-up condition. It is used to initialize the system and may be asserted asynchronously with the CLK/ signal.

PFW/ The power fail warning is an open collector, active low signal used to indicate an imminent power failure. It may be asserted asychronously with CLK/.

ID3/ through ID0/ These four active low, nonbussed signals are binary encoded and used to indicate a given board's position in the card cage.

NMRQ/ The active low nonmaster request line is used by boards not capable of becoming bus masters but which are in need of service.

7.2.3 Data Bus Signals

AD0/ through AD31/ These 32 tristate, active low address and data lines carry both address and data information. They are multiplexed so that the number of lines may be minimized. Address information is first broadcast. The responding slave then may place 8, 16, or 32 bits of data on these lines, depending on the type of transfer being performed.

TM0/ and TM1/ These two tristate, active low transfer mode lines are used to indicate the type of transfer being performed. The master initiating the transfer indicates the type of transfer to be performed and the responding slave indicates the success of the transfer over these lines.

ACK/ The acknowledge signal is an active low, tristate line used to indicate the completion of the data bus operation. ACK/ and START/, when used together, can signal that attention cycles have occurred on the bus.

START/ This active low, tristate signal is driven low to indicate the beginning of a data operation. ACK/ and START/, when used together, can signal an occurrence attention cycles on the bus.

SP/ The active low, tristate system parity line is driven to a state that produces even parity on the AD0/ through AD31/ lines. Parity may not be implemented on a given system. The status of the SPV/ line indicates whether this line has significance.

SPV/ This tristate, active low line is used to indicate the validity of the SP/ line. When driven low, the SP/ line must also be driven to a valid state.

7.2.4 Arbitration Bus Lines

RQST/ This open collector, active low line is used to indicate that a bus master requests use of the bus.

ARB0/ through ARB3/ These four open collector, active low lines are used to arbitrate among multiple masters requesting use of the bus. They are binary coded. Each board compares its own binary code with that on the ARB0/ through ARB3/ lines. At the end of an arbitration cycle, these lines carry the binary code of the master that will control the bus.

7.3 Types of Bus Cycles and Transactions

There are several types of bus cycles supported by NuBus. A bus cycle occurs during the 100-ns period between two consecutive rising edges of the CLK/ signal. Note that all signals change on the rising edge of the clock and are guaranteed stable on the falling edge of the clock. The type of bus cycle being performed is determined by the state of the following six lines:

1. START/
2. ACK/
3. TM0/ and TM1/
4. AD0/ and AD1/

Start Cycle This cycle is signaled by

$$\text{START/} = 0$$
$$\text{ACK/} = 1$$

TM0/, TM1/, AD0/, and AD1/ are used as four bits and encode one of 16 possible bus operations. Eight are read operations and eight are write operations.

Attention Cycle This cycle is signaled by

$$\text{START/} = 0$$
$$\text{ACK/} = 0$$

TM0/ and TM1/ are then used by one of four possible types of attention cycles, only two of which are currently defined by NuBus specification.

Ack Cycle This cycle is signaled by

$$\text{START/} = 1$$
$$\text{ACK/} = 0$$

TM0/ and TM1/ are used to indicate one of four possible acknowledge codes. They are

1. Successful completion of operation
2. Error occurred during operation
3. Bus timeout error occurred
4. Operation postponed

All bus transactions must begin with a start cycle and conclude with an acknowledge (ACK) cycle, so they must contain at least two clock periods.

7.3.1 NuBus Data Transfers

The bus supports the transfer of bytes (8 bits), words (16 bits), and long words (32 bits). The specification calls them bytes (8 bits), half-words (16 bits), and words (32 bits). There are four types of byte transfers. Byte 0 is always transferred on AD0/ through AD7/. Byte 1 is always transferred on AD8/ through AD15/. Byte 2 transfers are done on AD16/ through AD23/ and byte 3 transfers are performed only on AD24/ through AD31/. In a similar manner, there are two word transfer types. Word 0 is transferred on AD0/ through AD15/ and word 1 is transferred on AD16/ through AD31/. A long-word transfer is performed on AD0/ through AD31/.

The type of data transfer is controlled by TM0/ and TM1/ in conjunction with AD0/ and AD1/. A summary of the data transfer type and state of these four lines is shown in Figure 7-3. Note that when TM0/, TM1/ = 00, a byte write operation is being performed with AD1/ and AD0/ being used for the byte address decode function. When TM0/, TM1/ = 10, a byte

Transfer types	Transfer mode lines		Address lines	
	TM1/	TM0/	AD1/	AD0/
Write byte 3	0	0	0	0
Write byte 2	0	0	0	1
Write byte 1	0	0	1	0
Write byte 0	0	0	1	1
Write word 1	0	1	0	0
Block write	0	1	0	1
Write word 0	0	1	1	0
Write long word	0	1	1	1
Read byte 3	1	0	0	0
Read byte 2	1	0	0	1
Read byte 1	1	0	1	0
Read byte 0	1	0	1	1
Read word 1	1	1	0	0
Block read	1	1	0	1
Read word 0	1	1	1	0
Read long word	1	1	1	1

Figure 7-3 NuBus data transfer summary.

7.3 Types of Bus Cycles and Transactions

Transfer response	Transfer mode lines	
	TM1/	TM0/
Normal completion	0	0
Error	0	1
Timeout error	1	0
Device busy	1	1

Figure 7-4 NuBus transfer completion status.

read operation is being performed with AD1/ and AD0/ still performing the byte address decode function. TM1/, TM0/ = 01 indicates a write operation other than byte write. It can mean word 0 or 1, long-word, or block write operation. When TM1/, TM0/ = 11, a read operation other than byte read is being performed.

Parity checking is an optional feature on NuBus. If used, the bus master must generate the SP/ and SPV/ signals. The SPV/ signal indicates that parity checking is in effect and is used to check both address and data parity. SP/ is driven such that there are an even number of asserted ADn/ lines. The SP/ line is driven valid, assuming that all 32 ADn/ lines are being used even if a word or byte transfer is being performed. If a parity error occurs during an address transfer, there will be no response by any slave. If parity error occurs during a read data transfer, the transfer will be completed but the master receiving the data must assume the data are lost. If a parity error occurs during a write data transfer, the selected slave generates an error code reply.

Any slave responding to a transfer drives the TM0/, TM1/ lines to indicate the status of the transfer. A summary of the responses is shown in Figure 7-4. The four transfer status types can indicate a normal completion, delay of transfer to a future time, or two error conditions. Delay of transfer means that the master trying to perform the operation should retry the transfer.

Read Transfers

The timing diagram for a NuBus read transfer is shown in Figure 7-5. The read operation begins with the current bus master driving the START/, ADn/ with address information, and TMn/ with transfer mode information lines active synchronously with the rising edge of the CLOCK/ signal. The slave decodes the address information and responds to the read request. This is done by driving ACK/ active low, indicating a response and placing data on ADn/ lines and status code on TMn/ lines. All data, status code, and acknowledge lines are driven active synchron-

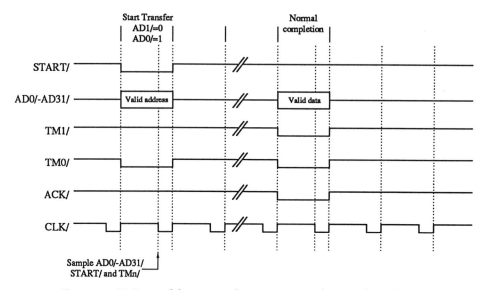

Figure 7-5 NuBus read–byte 2—read–operation example normal completion.

ously with the rising edge of CLOCK/. As in all types of transfers, address, data, status code, and acknowledge lines are stable on the falling edge of the clock. There may be more than one clock period between the start cycle and the acknowledge cycle.

Write Transfers

At the rising edge of the clock (CLK/) signal, the bus master drives the ADn/ with valid address data, TMn/ with mode data, and the START/ low to start the transfer. On the next clock cycle following the "START" cycle, the master drives valid data onto the ADn/ lines. The slave decodes the address information and takes the data from the ADn/ lines. The slave then completes the transfer by driving the status code onto TMn/ lines and driving the ACK/ line active low, indicating the transfer is completed. This process is shown in Figure 7-6.

Block Operations

Block operations are used when the bus master is performing an operation requiring the transfer of more than one long word. Data transfers are performed to sequential memory locations. As discussed earlier, read and write operations begin with a "START" cycle and end with an "ACKNOWLEDGE" cycle. Between these two cycles, several data transfers occur.

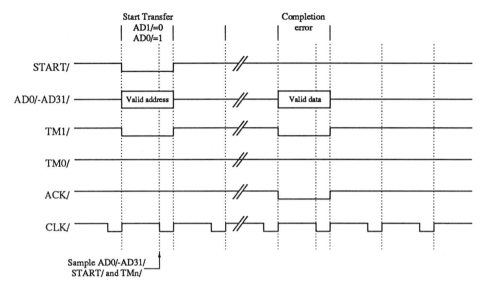

Figure 7-6 NuBus write long word, write operation example—completion error.

The size, or number, of transfers to be performed during the block transfer is determined during the start cycle. There are only four sizes of transfer permitted by NuBus. They are 2, 4, 8, and 16 long words, and only long-word transfers are permitted with block transfers. Lines AD5/ through AD2/ determine the number of transfers to be performed during the block operation. Note that only these codings are supported. Block size codes are summarized in Figure 7-7. Each data transfer after the start cycle is acknowledged by the slave by driving the TM0/ line low. TM1/ remains inactive high throughout the data transfer phase. Each data transfer may require more than one clock cycle. The clock cycle after the last data transfer terminates the block operation in a manner identical to a single transfer operation with the status code being driven. The timing diagram for a block read is shown in Figure 7-8 and for a block write in Figure 7-9.

In the event that a slave is unable to perform the block operation the master is requesting, the selected slave responds to the start cycle by issuing an acknowledge cycle without performing any data transfers. This is not considered an error but simply indicates that the slave is not capable of performing the block operation. In the event that an error occurs during the block transfer, the slave generates an acknowledge and drives TM1/ and TM0/ with the error code. All data are considered lost during any block transfer that ends in an error cycle. The slave may not indicate that an error occurred when the error occurs but may indicate an

Starting address	AD5/	AD4/	AD3/	AD2/	AD1/	AD0/	Block size
000000	0	0	0	1	0	1	
001000	0	0	1	1	0	1	
010000	0	1	0	1	0	1	
011000	0	1	1	1	0	1	2
100000	1	0	0	1	0	1	
101000	1	0	1	1	0	1	
110000	1	1	0	1	0	1	
111000	1	1	1	1	0	1	
001000	0	0	1	0	0	1	
011000	0	1	1	0	0	1	4
101000	1	0	1	0	0	1	
111000	1	1	1	0	0	1	
010000	0	1	0	0	0	1	8
110000	1	1	0	0	0	1	
100000	1	0	0	0	0	1	16

Figure 7-7 NuBus block transfer size codes.

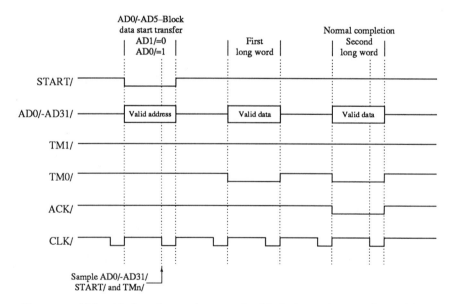

Figure 7-8 Nubus block read operation example, (block size = 2)—normal completion.

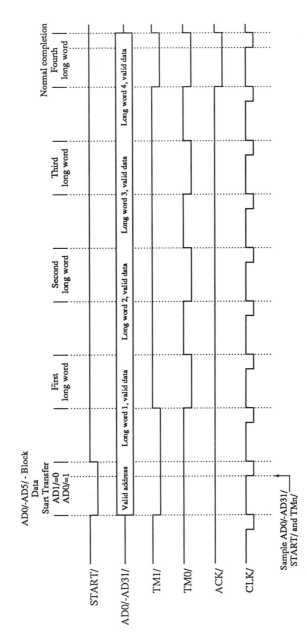

Figure 7-9 NuBus block write operation example (block size = 4)—normal completion.

error at any time after the error is detected up to and including the last acknowledge cycle.

7.4 Bus Arbitration

Arbitration is the process by which the master of the bus is determined. The bus master or "owner" initiates all bus operations. Like all busses, there are a specific set of lines that must be monitored and driven for every type of bus operation.

There may be as few as one or a number of bus masters. If there is one bus master, the arbitration process is simple. The master asserts RQST/ and the requester determines that it will gain control of the bus. For the case of several masters, several masters may be driving the RQST/ line simultaneously. The arbitration process awards control of the bus to a given master. This master assumes control of the bus immediately if the bus is not busy. If a transaction is in progress, the transaction is completed and the new master gains control of the bus on the clock cycle following the acknowledge cycle. Any master may assert the RQST/ line under the condition that it was not asserted during the previous falling edge of the clock. This provides for a "fair" bus arbitration scheme. Once a master has control of the bus, it remains the bus master until a new RQST/ is generated by another master and a new arbitration occurs. The current bus master may not begin another transaction if a RQST/ is generated by another master until the arbitration has decided the new master.

There are two attention cycles that control which master is currently in control of the bus.

7.4.1 Attention Operations Controlling Arbitration

An attention cycle occurs within one clock period and is generated by the bus master. It is indicated by the master driving both START/ and ACK/ active simultaneously. There are four types of attention cycles as follows:

TM1/	TM0/	Cycle Type
0	0	Null
1	0	Resource lock

Codes 0,1 and 1,1 are reserved and may not be used.

7.4 Bus Arbitration

The null cycle is used to force an arbitration cycle. This attention cycle may be used to indicate that a resource is no longer locked or that a bus master no longer needs the bus or that the current bus master no longer needs the bus and another master is requesting the use of the bus by driving the RQST/ line active low. Any null cycle forces a bus arbitration.

The resource lock cycle is used to fix or lock a device being used for a sequence of operations following the resource lock cycle. Note that the resource lock may not be "nested." There can be only one resource lock performed before the issuance of an "unlock" or null cycle to unlock the resource. The feature of "locking" resources is helpful when shared resources are used in a multimaster environment.

7.4.2 Arbitration Process

The bus lines used for the arbitration process are

RQST/ Active low open collector line used to indicate that a given master requests the bus.

ID3/–ID0/ Active low binary coded values that are assigned to each potential bus master.

ARB3/–ARB0/ Active low open collector lines driven by masters to determine the next bus master.

The arbitration process begins with a master desiring to use the bus driving the RQST/ line active low. This may be done only if the RQST/ line was inactive during the sample edge. All modules drive their unique IDs onto the ARB3/–ARB0/ lines. Each module continues to drive the RQST/ line active until it is granted control of the bus. The arbitration process may not take more than two clock cycles, at the end of which the ARBn/ lines contain the identification of the highest priority master. For example, assume master 14 (1110) is driving ARB3/–ARB0/ with the value 0001 and master 8 (1000) is driving ARB3/–ARB0/ with the binary value 0111. Since the lines are open collector, the final value on ARB3/–ARB0/ will be 0001 and master 14 will win the arbitration. The new master waits until the acknowledge cycle is complete and then takes control of the bus and starts performing bus transfers. At this point it also stops driving the RQST/ line. Note that it may not reassert RQST/ until all of the masters requesting the bus have gained control of the bus and the RQST/ line is inactive. This ensures the fairness algorithm or that every master will gain control of the bus. Note also that the new master may not wait more than 255 clock cycles before beginning the first transaction. Each master driving the ARBn/ lines will drive the lines with its unique identification code but must release the lines if a code higher than its own ID code is detected on the ARBn/ lines. Each slot has a unique IDn/ code assigned to that slot. The IDn/ lines are not bussed to each slot. The arbitration process is shown in Figure 7-10.

7.5 Geographic Addressing

The ID3/–ID0/ lines are used to assign a portion of the upper 256 Mbyte address space to each module plugged into a given slot. This represents only 6.2% of the 4096 Mbyte address space of NuBus. Since there are 16 codes assigned to IDn/ lines, each slot occupies 16 Mbytes with 16 slot codes for a total of 256 Mbytes. These addresses start at physical address F0000000 and extend to FFFFFFFF (hex). This process allows for the design of boards that have no configuration jumpers (if desired). The low address space (00000000 through 0FFFFFFF) is system-dependent, but the address space taken by a given board must be programmed by registers located in the upper 16 Mbyte space located in the geographic address space.

7.6 Utility Functions

Other lines not previously mentioned include

RESET/ An open collector, active low line used to reset all boards to their power-up condition. It must be asserted during power-up and power-down and may be asserted asynchronously with respect to the clock at any other time. The reset line is guaranteed active low for at least 100 Ms after the supply voltages have become stable following a power-up.

CLK/ The 10-MHz 25% duty cycle signal used to synchronize all data transfers and bus arbitration cycles. All bus signals are changed on the rising edge of the clock and are guaranteed stable on the falling edge of the clock.

PFW/ The power fail warning signal is an open collector, active low line used to indicate a power failure. It does not need to be asserted synchronously with the clock. This line is asserted at least 2 ms before the RESET/ line is asserted during a power failure. PFW/ must be inactive at least 1.0 ms before reset returns to its inactive level.

NMRQ/ The open collector, active low, nonmaster request signal is asserted for at least one clock period and is used to indicate that a board

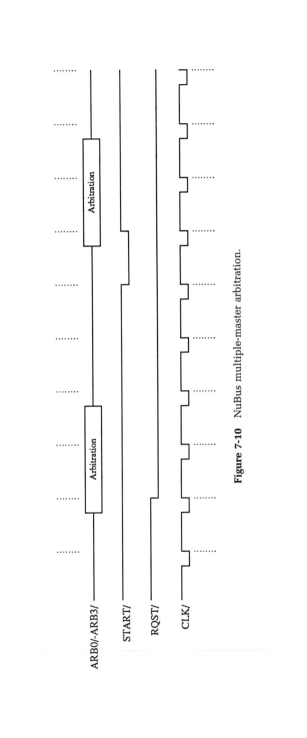

Figure 7-10 NuBus multiple-master arbitration.

requires some type of attention. The board continues to assert NMRQ/ until it has been serviced.

7.7 NuBus Electrical Characteristics

The voltages required to define a logic high or low at the driver and receiver are different. They are

	Voltage	
Line state	Receiver	Driver
High	>2.0 V	>3.5 V
Low	<0.8 V	Not specified

NuBus specifies the driver current handling requirements in terms of both AC and DC drive requirements in both the current sink and current source mode of operation. There is a pull-up and pull-down transient current as well as a steady state DC drive requirement on all lines. Those requirements are summarized in Figure 7-11. NuBus defines the transient

Line type	DC driver requirements		Transient driver requirements		Single load board	
	I (sink) (mA)	I (source) (mA)	I (sink) (mA)	I (source) (mA)	I (sink) (mA)	I (source) (mA)
Open collector	60		80		0.625	0.1
Address data	24	1.6	80	40	0.5	0.1
Control	24	1.6	80	40	0.5	0.1
Clock	60	30	90	50	1.4	0.1

Line type	Line characteristic impedance (Ohms)
Open collector	65
Address data	175
Control	175
Clock	65

Figure 7-11 NuBus driver and line characteristics.

7.7 NuBus Electrical Characteristics

period as ≤ 17.0 ns. This number is derived from the anticipated round-trip propagation delay for a typical backplane design with installed boards.

7.7.1 Power Supply Voltages

Four supply voltages are specified for NuBus. They are +5, +12, and −12. −5.2 is optionally specified but if it is not implemented the pin must be left unused.

Pin number	Row A	Row B	Row C
	P1 connector		
1	-12 VDC	-12 VDC	RESET/
2	RESERVED	RESERVED	RESERVED
3	SPV/	+5 VDC	+5 VDC
4	SP/	+5 VDC	+5 VDC
5	TM1/	+5 VDC	TM0/
6	AD1/	+5 VDC	AD0/
7	AD3/	-5.2 VDC	AD2/
8	AD5/	-5.2 VDC	AD4/
9	AD7/	-5.2 VDC	AD6/
10	AD9/	-5.2 VDC	AD8/
11	AD11/	GND	AD10/
12	AD13/	GND	AD12/
13	AD15/	GND	AD14/
14	AD17/	GND	AD16/
15	AD19/	GND	AD18/
16	AD21/	GND	AD20/
17	AD23/	GND	AD22/
18	AD25/	GND	AD24/
19	AD 27/	GND	AD26/
20	AD29/	GND	AD28/
21	AD31/	GND	AD30/
22	GND	GND	GND
23	GND	GND	PFW/
24	ARB1/	-5.2 VDC	ARB0/
25	ARB3/	-5.2 VDC	ARB2/
26	ID1/	-5.2 VDC	ID0/
27	ID3/	-5.2 VDC	ID2/
28	ACK/	+5 VDC	START/
29	+5 VDC	+5 VDC	+5 VDC
30	ROST/	GND	+5 VDC
31	NMRQ/	GND	GND
32	+12 VDC	+12 VDC	CLK/

Figure 7-12 NuBus P1 connector pin assignment.

7.7.2 Pin Assignment

Figure 7-12 summarizes the pin assignments for the P1 connector. Figure 7-13 shows the pin assignment for the P2 and P3 connector used on the triple-height-style board.

Pin number	P2 and P3 connector		
	Row A	Row B	Row C
1	User defined	User defined	User defined
2		GND	
3		GND	
4		User defined	
5		+5 VDC	
6		+5 VDC	
7		+5 VDC	
8		User defined	
9		User defined	
10		-5.2 ENAB	
11		-5.2 OUT	
12		GND	
13		User defined	
14		+12 ENAB	
15		+12 OUT	
16		GND	
17		User defined	
18		User defined	
19		GND	
20		-12 OUT	
21		-12 ENAB	
22		User defined	
23		GND	
24		User defined	
25		User defined	
26		User defined	
27		User defined	
28		+5 VDC	
29		User defined	
30		GND	
31		GND	
32	User defined	User defined	User defined

Figure 7-13 NuBus P2 and P3 connector pin assignment.

8

PC/XT/AT Bus

8.0 PC/XT/AT Bus Overview

These busses, all similar, were originally defined and developed by IBM Corporation for use in their PC/XT and AT series of microcomputers. This series of PCs was among the first truly mass-produced machines that brought the idea of a home computer into many homes and offices. Before this development, in the early 1980s, computers had been largely relegated to large businesses, universities, and home computer fanatics.

Except for minor differences, the PC and XT busses are identical. The AT bus expands the addressing space of the bus, increases the number of interrupt request lines and increases the width of the data bus. These changes were made to accommodate the increased capability of the Intel 80286 (CPU) which was used in the original IBM AT and which replaced the 8088 used in the PC and XT machines. The AT bus is a superset of the earlier PC/XT bus in that boards designed for the PC/XT can usually be used in an AT bus. However, AT bus boards may not be used in a PC/XT bus. Current versions of the AT bus run at far greater speeds than older machines. Most machines currently use the AT type of architecture; however, there is still a strong market for the original simple PC/XT machines. The newest machines use the 80386 and even the 80486 processors, which are a step up in power and speed over the 80286.

Since each processor in this family hosts an instruction set that is a superset of its predecessors, software developed for the original PC/XT machines will (usually) run on even the newest machines in the PC family. Problems can occur in some software that uses software delay loops for timing. Although this practice has been largely eliminated in the past few years, the problem still appears occasionally. As machine speeds increase, software delays decrease, causing errors. There can also be some problems with trying to run currently written software under old versions of DOS. DOS, or disk operating system, is the most common, but certainly not the only, operating system used on the PC family of machines. There have been a number of changes and upgrades to DOS since its earliest releases, resulting in compatibility problems between newly written software running on old versions of DOS. Most software will have a warning in the installation instruction regarding the oldest version of DOS that is supported by the software.

This series of busses are not true busses in the narrow definition of the word. These busses are designed around a specific processor family rather than a universal architecture. The CPU and support devices and some RAM and ROM are (usually) located on the motherboard or backplane rather than on plugs in daughter boards. The speed with which the bus operates depends upon the speed of the processor rather than being independently specified. Unlike most busses, this bus operates synchronously, using a master clock to time all transfer signals. The specific drive characteristics of lines are controlled by convention and specific implementation rather than specification.

8.1 Basic Features and Capabilities of PC Busses

The features supported by the PC busses include

	PC/XT	AT
1. Data bus width	8	16
2. Address bus width	20	24
3. Number of bus masters	1	1
4. Data transfer rate	Processor-dependent	
5. Bus error detection	Yes	Yes
6. Several mechanical sizes of boards		

Although the addressing range of the PC is 1 Mbyte and the AT is 16 Mbytes, DOS uses all of the memory from 640 kbyte to 1 Mbyte for BIOS (basic input/output system), Basic, and other system functions which are all based in ROM. The 80286-based AT has 16 Mbytes of addressing space but uses only the low 640k for many applications in order to maintain software compatibility with the older machines. Some software has been written to operate the 80286 in protected mode, which allows the machine to make use of the full 16 Mbytes of memory. Memory above the 1-Mbyte boundary is usually called "extended" memory. There is a standard that was originally developed and supported by a consortium of Lotus, Intel, and Microsoft which uses a paged memory scheme to make the machine appear to have a memory much larger than the basic 640k. This LIM standard is used by a number of major software developers. When this type of memory expansion scheme is used, it is called "expansion" memory. Each type of memory must be specifically supported by the software that is being operated on the machine. The vast majority of programs designed for use on the PC-based machines use only the low 640k of memory and do not use either expansion or extended memory.

The width of the data bus can be a confusing issue. The PC/XT bus uses a multiplexed data bus. The 8088 uses an external 8-bit data bus but an internal 16-bit architecture. Performing a 16-bit transfer is done by performing two 8-bit transfers. The high byte is transferred on the first part of the cycle and the low byte on the second part. The AT (80286)-based machines can also support this multiplexed data bus type of operation but additionally can support full 16-bit transfers.

8.2 PC Bus Mechanical Specifications

There are two large-board mechanical outline specifications for the PC and the PC/AT. Additionally, there is a small-board standard that is used to fit in the two end slots. The available space in these slots is reduced because of interference with the disk drive. Boards designed to operate in the PC or XT will also operate in the AT in many cases. In some instances, the PC or XT boards will not operate fast enough for the increased speeds used in newer machines.

The outline drawing of the full-sized PC and XT board is shown in Figure 8-1. The outline drawing of the PC and XT reduced-size board is shown in Figure 8-2. The outline drawing of the AT board is shown in Figure 8-3.

Note the presence of two connectors on the AT board. The larger 62-pin connector used on the PC and XT bus is identical to the 62-pin connector on the AT bus, both mechanically and electrically. The pins are labeled A1 through A31 and B1 through B31. The "A" row is on the component side of the board and the "B" row on the clad side.

The small PC and XT board basic dimensions are 3.90 in. in height and 6.00 in. in length. The basic dimensions of the full-sized PC and XT

8. PC/XT/AT Bus

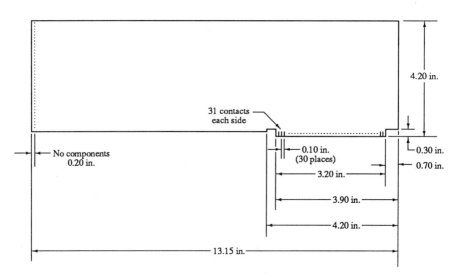

Figure 8-1 Full-size PC/XT board outline.

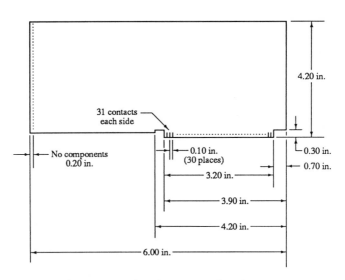

Figure 8-2 Reduced-size PC/XT board outline.

8.3 Bus Lines

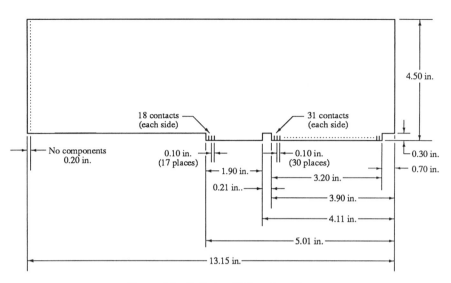

Figure 8-3 Full-size AT board outline.

board are 3.90 in. in height and 13.15 in. in length. The basic dimension of the AT board is 4.20 in. high and 13.15 in. long. The thickness of all boards is 0.062 in. Most boards also have a metal alignment bracket mounted on the end of the board nearest the connector. This bracket is also used for connector penetration, allowing connection of peripherals to the machine.

8.3 Bus Lines

The lines used on the PC and XT are the same. These lines are also used on the large connector of the AT board. The lines common to the PC, XT, and AT are defined in Section 8.3.1. The additional lines used by AT machines are defined in Section 8.3.2.

8.3.1 PC, XT, and AT Bus Lines

The pin assignments for the common lines are shown in Figure 8-4. The pin assignments for the AT bus unique signals carried on the 36-pin connector are shown in Figure 8-5. The lines are defined as follows:

GND Ground is carried on three pins and is used for power supply and signal return.

Component side				Clad side		
Pin number	Signal	Pin number	Signal	Pin number	Signal	Pin number	Signal
A1	IO CH CK/	A17	A14	B1	GND	B17	DACKA/
A2	D7	A18	A13	B2	RESET DRV	B18	DRQ1
A3	D5	A19	A12	B3	+5 VDC	B19	DACK0/
A4	D5	A20	A11	B4	IRQ2	B20	CLOCK
A5	D4	A21	A10	B5	-5 VDC	B21	IRQ7
A6	D3	A22	A9	B6	DRQ2	B22	IRQ6
A7	D2	A23	A8	B7	-12 VDC	B23	IRQ5
A8	D1	A24	A7	B8	CARD SLCTD/	B24	IRQ4
A9	D0	A25	A6	B9	+12 VDC	B25	IRQ3
A10	IO CH RDY	A26	A5	B10	GND	B26	DACK2/
A11	AEN	A27	A4	B11	MEMW/	B27	T/C
A12	A19	A28	A3	B12	MEMR/	B28	ALE
A13	A18	A29	A2	B13	IOW/	B29	+5 VDC
A14	A17	A30	A1	B14	IOR/	B30	OSC
A15	A16	A31	A0	B15	DACK3/	B31	GND
A16	A15			B16	DRQ3		

Figure 8-4 Pin assignment—PC/XT and AT bus common signals.

8.3 Bus Lines

Component side		Clad side	
Pin number	Signal	Pin number	Signal
C1	SBHE	D1	MEM CS16/
C2	LA23	D2	IO CS16/
C3	LA22	D3	IRQ10
C4	LA21	D4	IRQ11
C5	LA20	D5	IRQ12
C6	LA19	D6	IRQ15
C7	LA18	D7	IRQ14
C8	LA17	D8	DACK0/
C9	MEMR/	D9	DRQ0
C10	MEMW/	D10	DACK5/
C11	SD08	D11	DRQ5
C12	SD09	D12	DACK6/
C13	SD10	D13	DRQ6
C14	SD11	D14	DACK7/
C15	SD12	D15	DRQ7
C16	SD13	D16	+5 VDC
C17	SD14	D17	MASTER/
C18	SD15	D18	GND

Figure 8-5 Pin assignment—AT bus unique signals.

+5 +5 VDC power is carried on two pins. It is used to power devices on the slave board.

−5 −5 VDC power is usually no longer used on most new implementations of the PC. It was originally carried on one pin.

+12 +12 VDC power is carried on one pin. It is used to power analog devices or serial port (RS-232) devices using these higher logic levels.

−12 −12 VDC power is carried on one pin. It is used in a manner similar to +12 VDC.

OSC The oscillator pin carries a 14.318 MHz, 50% duty cycle clock signal. This signal is a general-purpose signal synchronous with the clock signal. The frequency of this signal varies with implementation.

CLK This 4.77-MHz, 33% duty cycle clock signal is used to synchronize all bus transfers. This signal is used to generate the basic "T" cycles used by the CPU to control the transfer of all data. The original PCs used a 4.77-MHz clock signal but this signal is implementation-dependent. The original ATs used a 6-MHz clock.

Reset Drv This active low reset signal is used to reset the machine to a known state. This occurs whenever the machine is first powered up. It may also be driven low at any time to produce a system reset.

A0–A19 These lines are 20 tristate, active high address lines. The addressing range of the bus is 1 Mbyte. These lines may be driven by the processor (motherboard) or a DMA controller.

D0–D7 These eight lines are active high, tristate, and bidirectional data lines. These lines may be driven by a DMA controller or the processor. Note that the width of the bus is 8 bits but the architecture of the 8088 is 16 bits internally. All 16-bit transfers to the processor are performed at two 8-bit transfers. This is referred to as a "multiplexed" data bus.

ALE This active low, tristate line is used by a slave board to load valid address information into address latches. It is called the "address latch enable" line and the falling edge of this signal is used to latch the data.

I/O CH Ck A low on this line indicates that a parity error has occurred. The width of the data bus is 8 bits, but a ninth bit has been added for parity checking.

I/O CH RDY This active low line is used to introduce wait states into the processor. It is sampled by the processor during the T2 cycle of each transfer. A low during this cycle causes the processor to wait one cycle and sample the state of this line again. The processor keeps inserting wait cycles until the line returns high.

IRQ2 through IRQ7 These active low lines are used to generate an interrupt request. IRQ2 has the highest priority.

IOR/ An active low on this line produces an I/O read command. This means that the mother board (CPU) is requesting a data transfer from the I/O device to the processor.

IOW/ An active low on this line produces an I/O write operation. The data is transferred from the motherboard (CPU) to the external I/O device.

MEMR/ (SMEMR/) An active low on this line produces a memory read command. Data is transferred from an external memory device to the CPU. This signal is called SMEMR/ on AT machines.

MEMW/ (SMEMW/) An active low on this line produces a memory write command. Data is transferred from the CPU to an external memory device. This signal is called SMEMW/ on AT machines.

DRQ1 through DRQ3 These are direct memory access (DMA) request lines 1 through 3. They are asynchronous channel requests used by peripheral devices to gain DMA services. DRQ1 is the highest priority. All DMA request lines are active high. When a DMA request line is drive

8.3 Bus Lines

active high, it is held high until the corresponding DACK line responds by being driven low.

DACK0 through DACK3 The lines are the DMA acknowledge lines corresponding to the DMA request lines. They are active low and are used to notify a requesting DMA device that it has been granted control of the bus so a DMA transfer can occur. In the PC, XT, and AT, it is also used to produce a dynamic memory refresh cycle.

AEN The address enable line is used to disable the processor and peripheral devices from using the bus. This is used to permit DMA transfers without bus contention. When active high, the DMA device has control of the address, data, and command lines.

T/C This active high signal is called "terminal count" and generates a pulse when the last data transfer during a DMA operation is reached. It notifies the CPU that it may resume control of the bus.

CARD SELTED A special-purpose signal used to indicate what is activated by cards in the short J8 socket. It is an active low signal.

Refresh/ This signal is driven active during a dynamic RAM refresh cycle. It is an active low TTL output.

OWS The active high signal is generated by the responding slave and indicates that the present bus cycle may be completed without any additional wait states.

8.3.2 Lines Specific to the AT

The following lines are implemented on AT-type machines only. All of these signals are carried on the smaller connectors or J10 through J14 and J16. Note that the PC/XT and AT common signals are implemented on J1 through J8. Newer machines use different motherboard layouts. The pin numbering for the 62-pin J1 through J8 connector is shown in Figure 8-6 and the pin numbering for the 36-pin J11 through J14 and J16 is shown in Figure 8-7.

SBHE The system bus high enable line indicates that the system is performing a 16-bit transfer. This is an active high, tristate signal.

LA23 through LA17 Unlatched address, active high lines that form the upper seven bits of the address bus. These lines, in conjunction with the lower address lines carried on the 62-pin connector, give the 16-Mbyte addressing range of the AT. They are valid while BALE is high and should be latched with the falling edge of BALE. They are tristate lines.

MEMR/ This active low, tristate signal is used to indicate that a memory read is commanded by the master. Note that SMEMR is driven

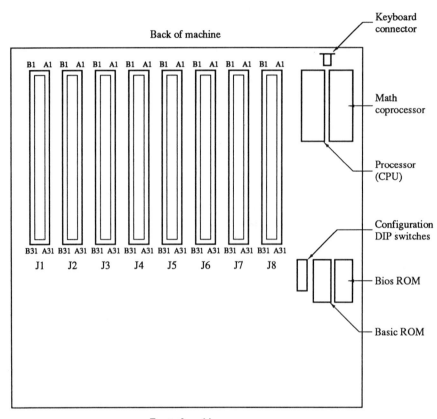

Figure 8-6 PC/XT and AT J1 through J8 pin identification.

active when the low 1 Mbyte is being addressed and both SMEMR and MEMR are driven active when addressing in the range from 1 to 16 Mbytes.

MEMW/ This active low, tristate signal is used to indicate that a memory read is commanded by the master. Note that SMEMW is driven active when the low 1 Mbyte is being addressed and both SMEMW and MEMW are driven active when addressing in the range from 1 to 16 Mbytes.

SD08 through SD15 Since AT machines are 16 bits, these form the upper 8 bits of the data bus. SD08 is the least significant bit of this upper byte. These are tristate, active high signals.

MEM CS16/ An active low, open collector or tristate signal used to indicate that the bus transfer is to be a 16-bit, 1 wait state memory access.

8.3 Bus Lines

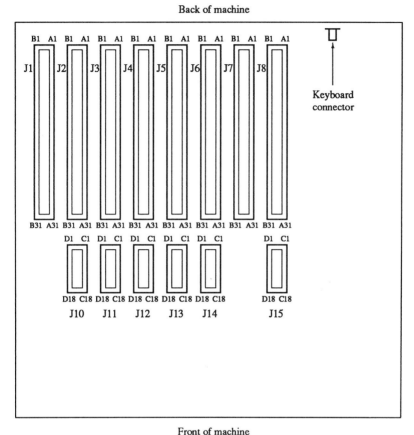

Figure 8-7 AT J1 through J16 pin identification.

I/O CS16/ An active low, open collector or tristate signal used to indicate that the bus transfer is to be a 16-bit, 1 wait state I/O device access.

IRQ10 through IRQ14 The lines are similar to the IRQ lines on the 62-pin connector. These four additional IRQ lines have a higher priority than IRQ3 through IRQ7.

MASTER/ This signal allows the system to support multiple masters, which are not supported in the PC or XT. It is used in conjunction with a DRQ lines. The bus master trying to gain control of the bus uses a normal DMA request cycle in which the DRQ line is driven active. When the DACK line echoes back active, the new master drives MASTER low. After MASTER is drive active, the new master may drive address and data lines valid after one clock period and the read and write command lines active after two clock periods. The new master may not control the bus for more than 15 µs or memory may be lost because memory refresh cycles are suspended while MASTER is active.

DRQ0, DRQ5 through DRQ7   These expanded DMA request lines work identically to the DMA request lines specific to the PC and XT.

DACK0/, DACK5/ through DACK7/   These expanded DMA acknowledge lines work identically to the DMA acknowledge lines specific to the PC and XT.

8.4 PC, XT, and AT Bus Cycles

There are four types of bus operations permitted on the PC bus:

1. Memory or I/O read operations
2. Memory or I/O write operations
3. DMA operations
4. Interrupt operations.

All memory or I/O operations are initiated by the motherboard where the heart of the machine resides. The motherboard, unlike most busses, contains the CPU, math coprocessor, memory, bus interface logic, realtime clock, and other functions. The bus contains eight expansion slots, six of which will accommodate the large or small outline board and two slots that will accommodate only small outline cards.

8.4.1 Memory or I/O Read Operations

The timing diagram for a memory read operation is shown in Figure 8-8. Note than an I/O read operation is nearly identical to a memory read operation. One wait state is inserted for I/O operations. On most newer (faster) machines, there are user options for adding additional wait states.

The PC, XT, and AT bus implementation uses a synchronous transfer, unlike many other busses. The transfers occur over four clock cycles. The transfer starts when the ALE signal is driven high during the low period of the clock. During this period, new address data are placed on the address bus and become stable before the falling edge of ALE. Address information may then be latched by the slaves on the falling edge of ALE. On the next falling edge of the clock after ALE goes low, the I/O or memory read command is driven low, depending on the type of read transfer. The memory or I/O device selected then has two clock periods to place valid

8.4 PC, XT, and AT Bus Cycles

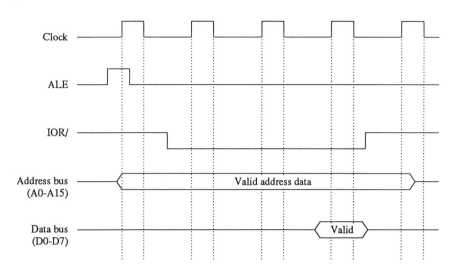

Figure 8-8 PC/XT bus memory read cycle.

data on the data lines. The master will latch the data and release the command line on the next falling clock edge. The slave driving data onto the data lines must then release the data bus lines within one clock cycle after this falling edge and the processor may then generate another ALE.

8.4.2 Memory or I/O Write Operations

The timing diagram for a memory write operation is shown in Figure 8-9. As in the case of the I/O read operation, an I/O write operation inserts one wait state. Other than this wait state, the operation is performed the same as a memory write operation.

The timing for write operations is similar to that for read operations. It also takes four clock cycles and is performed synchronously. The process starts when the ALE signal is driven high by the processor synchronously with the falling edge of the clock. This signal is held high for approximately one-half clock period during which new address data is driven onto the bus. This new address data may be latched by slaves on the falling edge of ALE. On the next clock falling edge after the ALE falling edge, the MEMW/ or IOW/ signal is driven active low, depending on the type of transfer being performed. This signal is held active for two clock periods. Synchronous with the falling edge of MEMW/ or IOW/, valid data is driven onto the data bus. This data may be taken by the receiving slave at the rising edge of the command line (MEMW/ or IOW/).

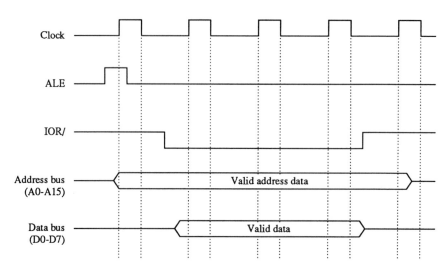

Figure 8-9 PC/XT bus memory write cycle.

8.4.3 DMA Operations

There are two types of DMA operations that may be performed by the bus. The timing diagram for a DMA read operation is shown in Figure 8-10 and the timing diagram for a DMA write operation is shown in Figure 8-11.

Like all bus transfers, DMA transfers are done synchronously with the clock.

DMA Read Operation

The device wishing to perform a DMA transfer initiates the process by driving the DRQ line active high synchronous with the falling edge of the clock. The motherboard arbitrates the DMA requests and returns an active low on the highest priority DACK line. The falling edge of the DACK line is synchronous with the falling edge of the clock. The DMA request is removed when the DMA device detects an active signal on its DACK line. The AEN line is driven active high, indicating that a DMA transfer is about to occur. Once the DACK line is driven low, the DMA device may drive valid address information onto the bus and begin transferring data. These transfers are performed using the same transfer format as a normal read operation where the IOR or MEMR lines are drive low to indicate the type of transfer.

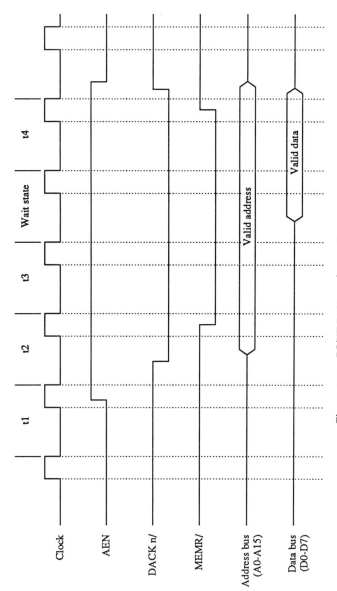

Figure 8-10 PC/XT DMA read operation memory.

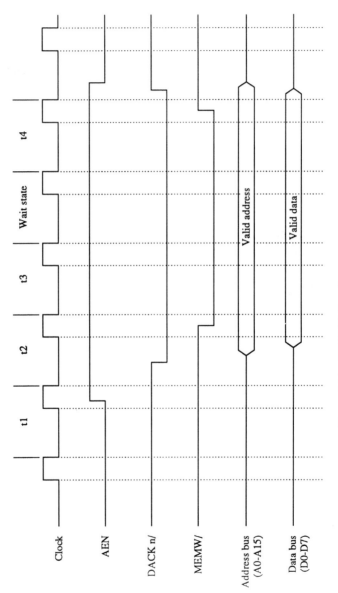

Figure 8-11 PC/XT DMA write operation memory.

DMA Write Operation

A DMA write operation is performed in a manner very similar to that of a read cycle. The DRQ line is driven active high by the DMA device. The motherboard drives AEN high, indicating that a DMA transfer is about to take place and selects the highest priority DMA request pending. The motherboard then drives the corresponding DACK line low. The DMA device then begins driving valid address data onto the data bus performing write operations like a normal memory or I/O write.

8.4.4 Interrupt Operations

The 8088, 80286, and 80386 have two generic types of interrupts. There are maskable and nonmaskable interrupts (NMI). The PC/XT and AT machines can receive an NMI from any of three sources. They are

1. RAM parity error
2. Coprocessor
3. I/O channel check request

Although these interrupts may not be masked (disabled) internal to the processor, the PC implementation allows for register masking of these interrupts.

In addition to the NMI, there are eight other interrupt lines, of which six are available on the bus. Of these six, three have been preassigned. The maskable interrupts are used as follows:

IRQ	Utilization
0	System timer, channel 0 (DRAM)
1	Keyboard interrupt
2	Not assigned
3	Not assigned
4	RS232 serial port
5	Not assigned
6	Disk drive
7	Parallel port

Note that only IRQ2 through 7 are available on the bus. The NMI has the highest priority and IRQ0 has the highest priority of the maskable interrupts. Since an 8259 interrupt controller is traditionally used in the PC, the priority of each interrupt may be rescheduled in software.

The interrupt process begins when a slave generates an interrupt request by setting its IRQ line to an active low state. The programmable interrupt controller (PIC), which is usually an 8259, receives the request and performs a priority arbitration. Assuming that the interrupt is not masked by the 8259 and it is the highest priority or only interrupt request pending, the 8259 generates an interrupt request to the processor. The processor responds by generating interrupt acknowledge level changes to the 8259. The first causes the current prioritization to be held constant and the second causes the 8259 to place an 8-bit pointer on the processor data bus. This pointer is used to form an address into the interrupt vector table in low memory of the CPU. The address of the entry point of the interrupt service routine is stored in the interrupt vector table. The complete address of the CPU is formed from the program counter (PC) and the code segment (CS), both of which are stored in the interrupt vector table. Both are 16-bit values and are added with a 4-bit offset. If the PC has a hex value of PPPP and the code segment has a value of SSSS, the physical 20-bit address, noted as AAAAA, is formed by adding the PC and code segment as follows:

$$\frac{\begin{array}{c}\text{OPPPP}\\ \text{SSSSO}\end{array}}{\text{AAAAA}}$$

Note that storage of the PC and CS values requires four bytes. Therefore, each entry in the interrupt vector table requires four bytes. The process of forming the complete 20-bit physical address from the CS and PC values is performed transparently by the processor.

9

STD Bus

9.0 STD Bus Overview

STD bus was originally developed as a venture of both ProLog and Mostek Corporations. It was first released in 1978 and has been made freely available to any user without royalty. Due to variations in implementations, a manufacturers' group was formed in 1980 to control uniformity and ensure standardization. Subsequently, a working group was formed at IEEE with the intent of developing STD bus into a standard. As of this writing, the final release of the proposed IEEE-961 specification is still pending. The current level of the proposed specification is Revision D1.

STD bus is a popular, low-cost, efficient way of implementing industrial and control types of applications. Traditionally, high-end computational engines have not been implemented on STD bus. However, there have been some proposed modifications to the bus that would allow it to operate up to a 32-bit data word. As currently implemented, however, it is a small-geometry, 8-bit bus with the capability of supporting one permanent master and several temporary masters. The temporary masters are intended to perform only DMA-type transfers. STD bus supports up to 24 bits (16 Mbytes) of address space. It also multiplexes the upper eight address lines onto the data lines. Bus-vectored interrupts are also supported by STD bus.

The largest single disadvantage of STD bus is that its implementations are processor-dependent. There are several variations on the bus implementation allowed in the specification such that the system designer must work closely with the prospective board vendor to ensure compatibility among devices.

The types of processors that have traditionally been implemented on STD bus include

1. 8080
2. 8085
3. Z80
4. 6800
5. 6809
6. NSC800
7. 6502
8. 6805

and other types of 8-bit or 16-bit processors that make use of a multiplexed data bus.

9.1 STD Bus Mechanical Outline

STD bus defines a card size of 6.50 by 4.50 in. with a thickness of 0.062 in. Because there is a no clad area of 0.10 in. on each side and the bus card edge connector requires 0.40 in., the actual usable component area is 6.10 by 4.30 in. If the (optional) card edge ejector is used, additional board space is lost due to the ejector. If a card ejector is used, the specifications call for a single ejector to be installed. The nature of card edge connectors is that they allow for accidentally inserting the board into the card rack upside down unless the cards are keyed. Cards are keyed between pins 25 and 27 as viewed from the component side of the board or between pins 26 and 28 as viewed from the circuit side. All even-numbered pins are on the component side of the board and all odd-numbered pins on the circuit side. Additionally, cards may be keyed for any given geographic slot location according to individual system requirements. Geographic keys may not be installed between pins 26 and 28 (or 25 and 27) as this nullifies the directional keying requirement.

Board spacing in a card rack is specified as 0.500 in. minimum. To achieve this spacing, the component height is limited to 0.375 in. and lead length on the clad side of the board to 0.040 in. This produces a minimum card clearance of 0.10 in. The outline of the STD card is shown in Figure 9-1. The bus uses a 56-pin card edge connector with connector fingers on 0.125-in. centers.

9.2 STD Bus Signal Lines

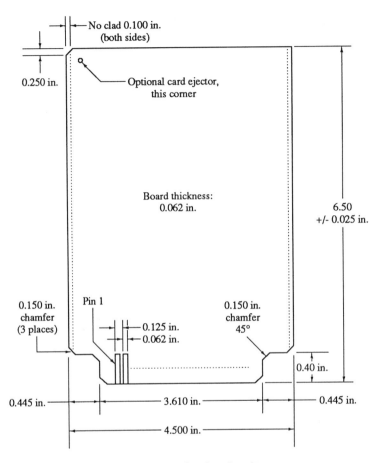

Figure 9-1 STD bus board outline.

9.2 STD Bus Signal Lines

STD bus is divided into five groups of lines. The groups are

- Power
- Auxiliary power
- Data
- Address
- Control

9.2.1 STD Bus Power and Auxiliary Power

There are up to five power supply voltages that may be supported by STD bus. The supply voltages are

1. +5 VDC: Carried on two pins
2. +12 VDC: Called auxiliary positive; carried on one pin
3. −12 VDC: Called auxiliary negative; carried on one pin
4. VBB#1/VBAT: Logic bias voltage or backup battery; carried on one pin
5. VBB#2: 5.0 VDC logic bias voltage; carried on one pin

Additionally, there are two pins dedicated to auxiliary power return and two pins dedicated to main power return.

9.2.2 Address and Data Lines

STD bus supports up to 24 address lines, but not all processor implementations support all 24 lines. Some devices, such as the 8080, Z80, and NSC800, utilize only 16 address lines and are not required to multiplex the upper address lines with the data lines. Other devices, such as the 8088 or 68008, support 20- to 24-bit addressing and must implement address/data multiplexing.

All eight data lines, referred to as D0 through D7, are active high, bidirectional, tristate lines. Data lines may be driven by the current master in the case of write operations or by the addressed slave in the case of read operations. All 24 address lines, referred to as A0 through A23, are tristate, active high lines. STD bus also supports I/O mapped devices. The low 16 bits are used to address I/O-mapped devices in the case of operations to I/O devices. The address bus is always driven by the current master. Note that since D0 through D7 and A17 through A23 are physically the same lines, these lines may be driven by a master only during the addressing portion of the transfer and may be driven by either the same master during a write operation or the addressed slave during a read operation.

9.2.3 STD Bus Control Lines

STD bus control lines may be broken into several functional groups. There are data transfer, interrupt, timing, and bus arbitration lines. There are a few utility lines not covered in the previous sections.

9.2 STD Bus Signal Lines

Data Transfer Control Lines

STD bus uses six lines to control the transfer of data. Those lines are

WR/ An active low, tristate signal used to indicate that the current master is performing a write to memory or I/O-mapped slave. This line may be driven only by the current master.

RD/ An active low, tristate signal used to indicate that the current master is performing a read from memory or I/O-mapped slave. This line may be driven only by the current master.

MEMRQ/ An active low, tri-state signal used to indicate that a memory (as opposed to I/O) transfer is being performed by the current master. This line may be driven only by the current master.

IORQ/ An active low, tristate signal used to indicate that an input/output (I/O) transfer is being performed by the current master. This line may be driven only by the current master.

MEMEX An active high signal driven from any source that enables memory overlays. When driven low, the "primary" memory is enabled. It is convenient to think of this line as an extra address line.

IOEXP An active high signal driven from any source that enables I/O overlays. When driven low, the "primary" I/O-mapped devices are enabled. As with MEMEX, this may be thought of as an extra I/O-mapped address line.

Interrupt Control Lines

INTRQ/ This line is a general-purpose, maskable, active low, open collector interrupt request line. Only one interrupt line is supported on STD bus. It may be driven by any interrupting slave. Note that only one permanent master may process interrupts.

NMIRQ/ This active low, open collector line is the nonmaskable interrupt line. This line would be used for very high priority types of interrupts such as a power failure. It may be driven by any interrupting slave or master.

INTAK/ This active low TTL signal is driven from the one permanent master and is used to indicate to the interrupting slave that the master is responding to an interrupt request. If there is more than one interrupting slave, a bus-vectored interrupt scheme must be implemented as there is only one interrupt line. The interrupt vector must be placed on the data lines during the time INTAK/ is active low.

STD Timing Lines

WAITRQ/ This active low open collector signal, when asserted, causes the current master to freeze the current state of the machine. This allows slow slaves to respond to faster masters. Since STD bus is synchronous, no handshake protocol is employed to ensure that a transfer was properly done. Most typically, this line inserts wait states into a processor.

CLOCK/ A TTL signal driven from the permanent master that is used to synchronize the system. It is the buffered processor clock signal.

CNTRL/ A general-purpose timing and control signal not specified to have any special characteristic. It is system-dependent. It may be a real-time clock signal or some multiple of the processor clock, for example.

REFRESH/ This active low tristate signal is driven by the master and indicates that a refresh cycle of the dynamic RAM is to occur. In those systems not using dynamic RAM, the line may be used for any general-purpose memory control function.

MCSYNC/ The machine cycle sync signal is an active low tristate signal used to define the start of the machine cycle. The function and timing of this signal are functions of the processor being used.

STATUS1/ A tristate, active low line used to indicate that an instruction fetch cycle is in progress by the processor. Not all processors implement this function. It may also be used as an auxiliary timing line.

STATUS0/ A tristate, active low line used as an auxiliary timing line. Its function varies with individual processors.

Bus Arbitration Lines

These lines are used in the process by which a temporary master may take control of the bus from the permanent master to perform a DMA transfer.

BUSRQ/ The bus request is an active low, open collector signal driven by a temporary master to notify the permanent master that a request to use the bus is pending.

BUSAK/ The bus acknowledge line is an active low, open collector signal driven by the permanent master. It is used to indicate that the bus is available for the requesting temporary master to use. Before issuing the BUSAK/ command, the permanent master completes the machine cycle it is currently executing. It tristates its drivers and only then issues the BUSAK/ command.

PCO/ The priority chain out is an active low TTL signal and is not bussed, but is daisy-chained with the PCI signal. This daisy chain is used when more than one temporary master is installed in the system. If a temporary master receives an active high on the PCI and is requesting the bus, it sends a low on this line (PCO) and the temporary master has control of the bus. If a temporary master receives a low (inactive) on PCI, it ripples out low regardless of its bus request status. If a temporary master receives an active high on PCI and is not requesting the bus, it ripples out a high on its PCO line. A serial arbitration scheme is implemented using this technique in which the priority of the temporary master is strictly slot-dependent.

PCI The priority chain in is an active high, nonbussed, TTL signal driven by all next higher priority temporary masters and the permanent master. It is used in conjunction with the PCO/ signal discussed previously.

Utility Lines

SYSRESET/ The system reset is an active low, open collector signal. It is used to initialize the system to a known state.

PBRESET/ The active low, open collector push-button reset signal is driven from any source capable of performing a system reset. It triggers a SYSRESET/ and is generally decoded at only one place.

The pin assignment for the STD bus is shown in Figure 9-2.

9.3 Data Transfer Operations

9.3.1 Compatibility

Card compatibility is not guaranteed by a manufacturer's statement that the card is IEEE-961 compliant. Card compatibility is determined by comparing the specified required access times of memory or I/O and the minimum required access time of the proposed bus master.

In the case of read operations, the time required for the slave to place stable data on the bus must be less than the minimum time required for the master to take the data. All master boards must specify the maximum time the board must require for read data to become stable after read command is issued. All slave boards must also specify the maximum required read access time or the time required for data to become stable after the read command is issued by the master.

In the case of write operations, the time required for the slave to accept data placed on the bus by the master must be less than the write cycle time of the master. All master boards must specify the minimum time required for data to be written to the slave to remain stable and the minimum time the data is held stable after the write command is inactive or the write hold time. All slave boards must specify the minimum time the data must be

Component side of board		Clad side of board	
Pin number	Signal function	Pin number	Signal function
1	+5 VDC	2	+5 VDC
3	GND	4	GND
5	Vbb1 / VBat	6	Vbb2
7	D3 / A19	8	D7 / A23
9	D2 / A18	10	D6 / A22
11	D1 / A17	12	D5 / A21
13	D0 / A16	14	D4 / A20
15	A7	16	A15
17	A6	18	A14
19	A5	20	A13
21	A4	22	A12
23	A3	24	A11
25	A2	26	A10
27	A1	28	A9
29	A0	30	A8
31	WR/	32	RD/
33	IORQ/	34	MEMRQ/
35	IOEXP	36	MEMEX/
37	REFRESH/	38	MCSYNC/
39	STATUS1/	40	STATUS0/
41	BUSAK/	42	BUSRQ/
43	INTAK/	44	INTRQ/
45	WAITRQ/	46	NMIRQ/
47	SYSRESET/	48	PBRESET/
49	CLOCK/	50	CNTRL/
51	PCO	52	PCI
53	AUX GND	54	AUX GND
55	AUX +V	56	AUX -V

Figure 9-2 STD bus pin assignments.

held stable after the write command is issued and the minimum time the data must be held stable after the write is brought inactive.

9.3.2 Read and Write Operations

STD bus supports the reading and writing of data from memory or input/output. Data may be transferred from the primary to alternate memory or I/O space by using the MEMEX and IOEXP lines. The address-

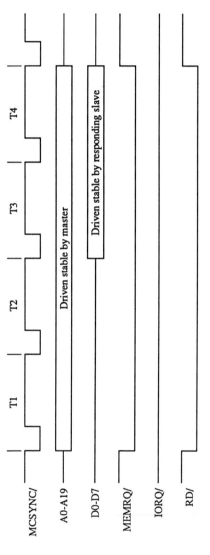

Figure 9-3 STD bus memory read operation.

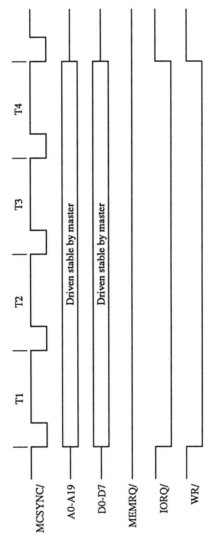

Figure 9-4 STD bus I/O write operation.

ing cycle begins with the processor driving the MEMEX or IOEXP signals valid concurrent with the address bus, MEMRQ/, or IORQ/ lines. The use of all of these lines selects the device to (or from) which the transfer is to occur. The RD/ signal is also driven valid in the case of a read operation. The responding slave places data on the bus, which is loaded by the master. A read operation is shown in Figure 9-3. This figure illustrates a memory read from primary address space.

In the case of a write operation, the WR/ signal is driven valid and data is placed on the bus concurrently with the WR/ signal being driven valid. A write operation is shown in Figure 9-4. This figure illustrates an I/O write operation to alternate I/O address space.

9.4 Signal Voltage Tolerances

STD bus has minimum and maximum values placed on signal voltages on all lines. The maximum voltage to be placed on any input is to be V_{cc} + 0.5 V, and the minimum voltage is to be −0.4 V. This ensures that devices will not be damaged. The tolerance on all supply voltages is ± 0.25 V with the exception of the auxiliary voltages which are ± 0.50 V. All drivers must deliver at least 2.4-V output at a load current of 15 mA and all receivers must accept any voltage above 2.0 V as a high state. All drivers must deliver less than 0.4 V at 24-mA load and all receivers must accept any voltage below 0.8 V as a low state.

10

Programmable Logic Devices

10.0 Programmable Logic Devices Overview

Although programmable logic devices, usually abbreviated PLDs, are not directly related to the design and implementation of microcomputer busses, they are a powerful tool that simplifies interface and on-board design tasks. They allow for flexibility in the initial design task, ease of debugging, and reduction of on-board pin count and real estate to implement any given function. Programming software is inexpensive and the parts themselves have improved considerably in performance and power in the past decade. During this time, they have also dropped in price to the point that nearly every complex board designed recently contains one or more of these devices.

Speeds for the parts range from combinatorial delays in the range of 5 to 7 ns and clock speeds for flip-flops ranging up to 100 MHz. Complexity ranges from 8 to 10 outputs and 8 to 10 inputs with a few dozen gates and fuses to upward of 75 combination input/output pins with upward of 10,000 fuses and several thousand gates. The internal architecture of the devices also varies tremendously from device to device with some architectures implementing a simple and/or combinatorial function to other devices consisting of arrays of and/or networks linked together with a hierarchy of local and global busses with each output containing a user-configurable macrocell. Parts may be one-time fuse-programmable or may be implemented with UV-erasable technology, allowing reprogramming of the parts. As would be expected, the more complex parts and the erasable parts tend to be architecturally more powerful but suffer from slower speeds and higher cost. As for any design decision, there is a cost performance trade-off that must be performed before selecting any part. Given the savings in board space, pin count, and increased density, PLDs are very often a cost-effective design option.

PLDs were developed to fill a specific design void that became apparent during the late 1960s and early 1970s. As digital design became more common, board-level designs had requirements for certain commonly recurring functions. There were memory-intensive functions, register-oriented functions, sequencing functions, and combinatorial functions, among others. A combinatorial device is a device whose output is an instantaneous function of current inputs. Real device outputs do not change instantaneously but have a characteristic "propagation" delay through each gate. This propagation delay determines the speed of the device. Combinatorial functions are implemented using AND, OR, NOT, and other types of gates, buffers, and inverters. the combinatorial PLD implements any arbitrary combinatorial function within the limitations of the size of the device.

10.1 Boolean Equation Review

This section provides a brief review of relevant digital design using Boolean equations as related to their utilization in PLDs. It is not complete and does not provide in-depth proof or demonstration of basic ideas. It does bring out important precepts that must be understood before a designer can use programmable logic devices effectively. If you are familiar with designing with Boolean equations, this section may be skipped.

Any combinatorial function can be represented by a Boolean equation. Regardless of the complexity of the equation, it can always be reduced to this basic sum of product term form. An example of such a Boolean equation might be

$$Z = (AB^*) + (AC) + B^*[(C^*D) + (A^*C^*D^*)] \qquad (10\text{-}1)$$

where the (+) terms represent a logical "OR" function and the terms inside the parentheses are the product terms and represent a logical "AND" function. Note that the equation has one output (Z) and four inputs (A,B,C,D). Also note that B^* is the inverted or complemented version of B. Complemented inputs are not considered to be unique inputs.

The combinatorial function may also be represented as a truth table, as shown in Table 10-1, or it may also be shown in Carnaugh map form as shown in Table 10-2. Using the Carnaugh map (or other techniques),

10.1 Boolean Equation Review

TABLE 10-1
Truth Table for Equation (10-1)

A	B	C	D	Z
0	0	0	0	1
0	0	0	1	0
0	0	1	0	0
0	0	1	1	0
0	1	0	0	0
0	1	0	1	0
0	1	1	0	0
0	1	1	1	0
1	0	0	0	1
1	0	0	1	1
1	0	1	0	1
1	0	1	1	1
1	1	0	0	0
1	1	0	1	0
1	1	1	0	1
1	1	1	1	1

it is possible to reduce the equation to a minimum form. A minimum form has the characteristic of having the least number of terms. The minimum form of Equation (10-1) is

$$Z = [AC] + [AB^*] + [B^*C^*D^*] \qquad (10\text{-}2)$$

As you can see, the initial form of the equation [Equation (10-1)] has four terms, while the reduced form has three. The original equation requires four INVERTERS, four AND gates, and two OR gates to implement as

TABLE 10-2
Carnaugh Map for Equation (10-1)

Z \ CD AB	00	01	11	10
00	1	0	0	0
01	0	0	0	0
11	0	0	1	1
10	1	1	1	1

$$Z = (AB^*) + (AC) + B^*[(C^*D) + (A^*C^*D^*)]$$

written. The reduced form of the equation requires three INVERTERS, three AND gates, and one OR gate. Reducing this equation has resulted in a savings of one inverter and two gates. Depending on the equation being reduced, the reduction process may result in an even more significant saving of "hardware." Another important feature of this equation is that it is implemented as a sum of products or the ORing of ANDed terms. Remember, all combinatorial functions can be reduced to this form.

Reduction of equations is an important feature of most PLD implementation software. Some software packages perform the process effectively, while others are less effective. All packages I have examined have weak points resulting in poor implementation, depending on the type of equations being reduced and the architecture of the part. Fortunately, most software designers also recognize the limitations of the algorithms and have incorporated provisions for working around the limitations.

The implication of having a poorly implemented equation reduction algorithm is that a given design may not be able to be implemented in a PLD of a given size. The options open to the designer are to increase the size of the PLD, reduce the equations more efficiently, or segment the design into multiple parts. The most cost-effective solution to the problem is to reduce the equations more efficiently. However, this is not always possible as there is a lower limit or optimal implementation.

10.2 State Machine Review

A second type of problem commonly implemented with a PLD is a state machine. A state machine is a "superset" of a combinatorial machine. The generalized block diagram of a state machine is shown in Figure 10-1. It consists of inputs, outputs, an input combinatorial network, a collection of storage elements, and feedback. Unlike the combinatorial machine in which the output is a function of current inputs only, the state machine produces outputs that are a function of current inputs and the past history of inputs. The storage elements implement the "memory" function.

State machines, as the name implies, are designed to progress through a succession of states in response to changing input conditions. There are two broad classes of state machines. The most common is the synchronous machine, in which the storage elements are clocked at regular intervals by a clock signal. A less common but much faster type of state machine is the asynchronous state machine. This type of state machine

10.2 State Machine Review

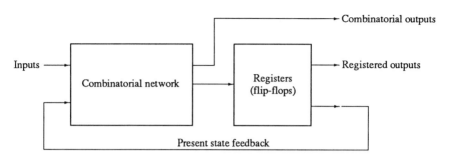

Figure 10-1 State machine block diagram.

relies on the inherent gate propagation delays through the combinatorial network to act as the storage elements. This type of design can be tricky to implement and especially tricky to debug.

By making use of the storage feature of state machines, it is possible to have multiple paths and decision-making ability based on previous inputs. The clearest representation of the operation of a state machine is a state machine flow diagram or "bubble chart." An example of a flow diagram is shown in Figure 10-2. The flow diagram shows each state the machine may be in, the state variables representing that state, and the conditions under which that state may be entered or left. The input conditions under which the machine moves from one state to another are called the "transition conditions."

Outputs can be generated in one of two ways. The equations may be implemented as a function of the current state of the machine, which is the easiest method, or they may be implemented as a function of the current state of the machine and the current inputs. The method of generating outputs by current state only results in an easy and simple implementation but suffers from the problem of the outputs reaching the desired stable state delayed one clock period longer than an implementation in which outputs are generated based on current state and current inputs. If the method of generating outputs by means of current state and current inputs is used, there is a high occurrence of "sneak pulses" appearing at the output due to two effects. Inputs which are noisy are more likely to produce short transient pulses on the output lines. Also, as the state machine transmits from one state to another, the state output may briefly enter an incorrect state lasting less than one clock period before settling into the correct state. The use of PLDs instead of discrete logic to implement state machines has greatly reduced the second source of "sneak pulses."

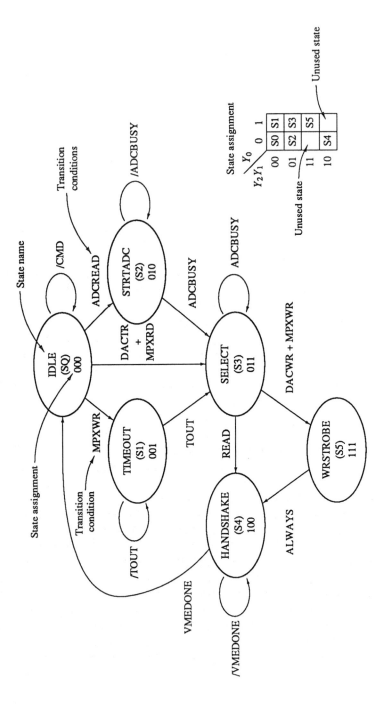

Figure 10-2 VME analog I/O board state machine flow diagram.

10.3 PLD Architecture

The following are a few selected examples of the internal architecture of some currently available devices. They were chosen to give a cross-sectional view of the variety and power of the parts. There are a large number of companies making PLDs in a wide variety of architectures. Each of the architectures has both strong and weak points in any given application. This section is not comprehensive and is included only to provide a brief overview of some parts currently available. The parts continue to improve in both speed and power with each passing month. Hardly a month goes by without a number of new parts being introduced.

10.3.1 Small-Scale Device (16L8)

Programmable logic devices' internal architectures and pin characteristics are variable. However, there are some universally used base architectures around which most devices are created. The earliest and most common of the architectures is called a PAL, or programmable array logic. An example of a simple and early, although still popular PLD of this type, is the 16L8. The internal model of the part is shown in Figure 10-3. The principal features of the 16L8 are a completely combinatorial architecture, 10 dedicated input pins, 2 dedicated output pins, and 6 bidirectional I/O pins. The bidirectional I/O pins can be used as inputs, outputs, or both, depending on the user's requirements. Note that the output buffers of bidirectional I/O pins are driven by a combinatorial term that can be derived from inputs and other outputs. The dedicated input pins are pins 1, 2, 3, 4, 5, 6, 7, 8, 9, and 11. The dedicated output pins are 19 and 12. The bidirectional I/O pins are 18, 17, 16, 15, 14, and 13. The configuration of the output stage consists of 7 AND gate outputs driving an OR gate which in turn is buffered by a tristate buffer forming the output signal. The tristate control signal is also driven by the input array. It is important to realize that the "schematic" representation of the 16L8 shown in Figure 10-3 is different from most schematics in that each AND gate shown driving the OR gate(s) is a 32-input gate. Since there are 10 input pins and 6 bidirectional I/O pins, there are a total of $2(10+6) = 32$ terms that can be used to generate any given output, which is why each AND gate is a 32-input gate. Note also that the 16L8 array has a large array of vertical lines. Each vertical line is connected to either an input, I/O signal, or the complement of an input or I/O signal. Each horizontal line coming from each AND gate is in

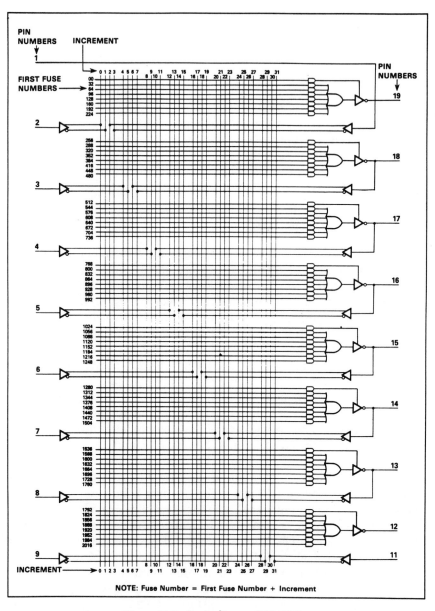

Figure 10-3 Logic diagram PAL16L8.

10.3 PLD Architecture

reality 32 lines, with each line connecting to one input, I/O, or complement line. Each horizontal line is connected to a vertical line by means of a fuse. When the device is shipped from the factory, all of the fuses are intact. The process of programming the device to implement a desire (combinatorial) function selectively "blows" unwanted fuses. Note that the device can be programmed only once, as a blown fuse cannot be repaired. Ultraviolet-erasable PLDs or EPLDs are similar to UV-erasable PROMs (EPROMs) in that the "fuses" may be erased by exposure to UV light.

To understand how this technique can be used to implement any combinatorial function, refer to Equation (10-2). This equation represents a sum of products reduced form of combinatorial function. The reproduced equation is

$$Z = (AC) + (AB^*) + (B^*C^*D^*) \qquad (10\text{-}2)$$

If this equation were to be implemented in discrete logic using AND, OR, and INVERT gates, the result would be as shown in Figure 10-4. the implementation takes the four inputs A, B, C, D or their complements and applies them to AND gates to form the product terms, which are (AC),

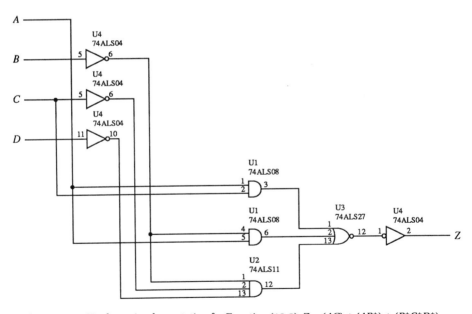

Figure 10-4 Hardware implementation for Equation (10-2): $Z = (AC) + (AB^*) + (B^*C^*D^*)$.

(AB^*), and ($B^*C^*D^*$). These three-product terms are then ORed to form the desired output. This circuit is identical to the architecture of one of the output pins (or I/O pin) of the 16L8. The differences are that two 2-input AND gates are required: one three-input AND gate, and one three-input OR gate are required to implement the function, while each output from the PLD provides for up to seven 32-input AND gates and a 7-input OR gate. This equation can be implemented using the 16L8 by "blowing" or disconnecting the unwanted input fuses from the unused inputs or unused AND gates using only three of the seven available AND gates and providing a permanent output enable to the tristate buffer at the output. As can be seen, the 16L8 has the limitation of not permitting the implementation of any combinatorial function that requires more than seven product (AND) terms or any single product term that requires more than 32 individual inputs and complements. Other devices allow more product terms and more inputs.

10.3.2 Medium-Scale Device (20RS10)

An example of a combinatorial PAL that supports more inputs and some other more powerful features is the 20RS10 shown in Figure 10-5. This device supports up to 40 input and complement terms and uses a shared OR output architecture. In this device, output pins or I/O pins are paired with each pair of pins, being limited to 14 product terms. Each output is also driven by an exclusive OR (XOR) gate. This feature implements a programmable inverter feature at the output. Equation (10-2) implemented the "ones" of the combinatorial function. In some instances, the number of product terms may be reduced by implementing the "zeros" of the function. In those instances the output of the OR gate has an inverted polarity over the desired output. By selectively blowing the fuse at one input to the XOR gate, the gate becomes an inverter. If the fuse is not blown, the gate becomes a noninverting buffer. By selecting whether the gate is inverting or noninverting, the output may implement the "zeros" or "ones" of the function, depending on which has the least number of product terms. An example of such a function is shown in Figure 10-6. This function can be implemented using the ones in the equation in four product terms. The same function can be implemented using the zeros of the equation in two product terms. It is more hardware-efficient to implement the zeros of the equation and then invert the output. The approach saves the use of two AND gates. For larger functions, this design variance may make the difference between a function fitting or not fitting in a given device.

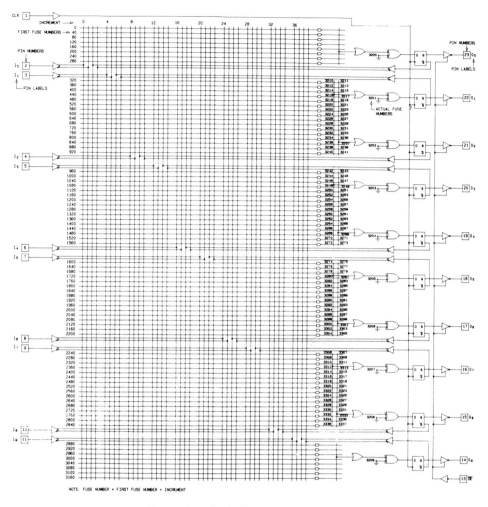

Figure 10-5 Logic diagram PAL20RS10.

| Z C,D | | | | |
A,B	00	01	11	10
00	1	0	1	1
01	1	0	1	1
11	1	0	1	1
10	0	0	0	0

Figure 10-6 Implementing zeros and ones of a Boolean equation. Equation implementing zeros: $Z = (AB^*) + (C^*D)$. Equation implementing ones: $Z = (A^*C) + (BC) + (BD^*) + (A^*D)$. Note that the implementation of zeros requires two product terms and the implementation of ones requires four.

10.3.3 Field-Programmable Logic Sequencer (FPLS) (82S105)

There are other types of PLDs that have an internal architecture optimized for state machines. An example of this is the device shown in Figure 10-7. The 82S105 is different in that it has both an AND and an OR fuse array. It also has eight outputs which are all registered and six

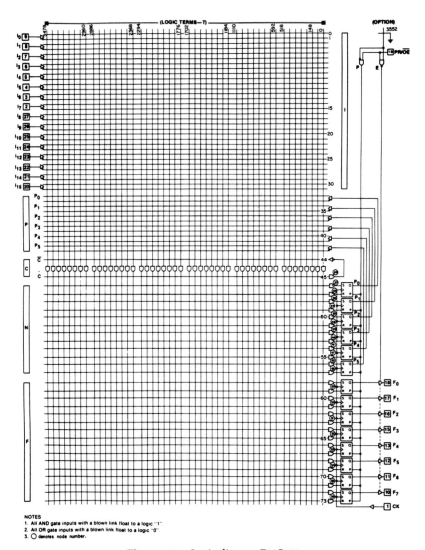

Figure 10-7 Logic diagram F82S105.

10.3 PLD Architecture

"buried" register outputs. A buried register is an output from the state machine that is not routed to a package pin. These buried registers are used for the machine state variables. The state of these six registers indicates the current state of the machine. Also note that output and state registers are clock SR flip-flops. This type of flop has the advantage of having relatively simple equations to implement the S and R inputs to the flop but it does require two sets of equations (S and R equations) for each state variable or output. One pin (19) can be programmed to be either a preset input (PR) or output enable (OE) input, depending on the state of fuse 3552. Note also that the device allows for 16 inputs in addition to the six state variable feedback terms and one nonregistered feedback term. This implies that each AND gate has 45 inputs. There are 48 of these 45-input AND gates feeding 29 OR gates, each having 48 inputs. This device is currently available with clock speeds up to 50 MHz.

10.3.4 Large-Scale Device (EP1800)

Another example of PLD architecture is one of the more powerful types of machines on the market today. As the complexity of the parts has increased, the number of gates and inputs for devices has increased. The EP1800 is marketed by Altera Corporation and consists of an array of PLDs housed in one package. The overall block diagram is shown in Figure 10-8. The EP1810 is an enhanced speed version of the EP1800 with a similar architecture. The device is split into four quadrants, with the four quadrants linked together by a global data bus. The global bus contains provisions for 12 global inputs, 4 clock or global inputs, and 4 of 12 macrocells in each of the four quadrants. There are therefore a total of 32 inputs or I/O lines and their complements carried on the global bus. Each of the four quadrants contains four macrocells. Each macrocell consists of a logic array, a configurable output cell, and an I/O buffer. See Figure 10-9 for a diagram of local and global macrocells. Each macrocell is equivalent to about 40 gates and the device to about 2000 gates. Of the 12 macrocells, four are global, having I/O pins that can be routed to the global bus, and the remainder are local to that macrocell. The outputs of each output cell may be fed back selectively to the macrocell logic array. The output macrocell may be configured into a number of convenient configurations including but not limited to

1. D flip-flop output, no feedback
2. Combinatorial output, no feedback
3. T flip-flop output, no feedback
4. SR flip-flop output, no feedback
5. D flip-flop output, feedback

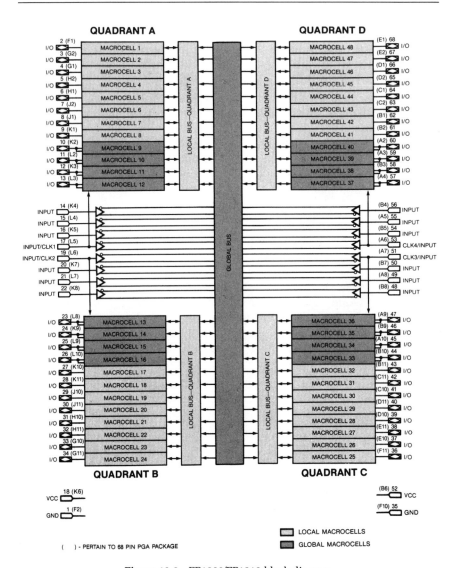

Figure 10-8 EP1800/EP1810 block diagram.

10.3 PLD Architecture

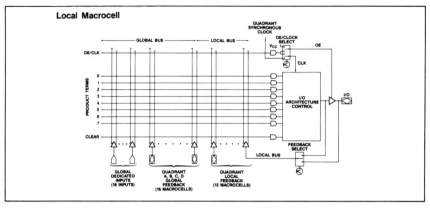

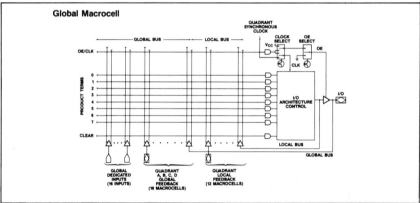

Figure 10-9 EP1800/EP1810 local and global macrocells.

6. Combinatorial output, feedback
7. T flip-flop output, feedback
8. SR flip-flop output, feedback
9. Others too numerous to list here . . .

The total number of configuration options for the output cell is large. Each logic macrocell has eight AND gates with inputs that can be derived from both the global bus and the local I/O pins local to that quadrant. This type of architecture can be used to implement both combinatorial and registed outputs such as are needed in state machine implementations. The device is also large enough so that several state machines that are independently clocked may be included in a single part along with the necessary decoding logic to generate the outputs. Unfortunately, as the devices get larger, speed suffers and this is particularly true of EPLDs. Small-scale parts like the 16L8 can currently be purchased that have delays as low as 7 ns, while parts like the EP1810 have delays for input to combinatorial output on the order of 30 ns. UV-erasable parts are at a definite speed disadvantage to fuse-programmable parts; however, they do have the advantage of being reusable. In a small production environment or during product development, this feature is very useful. Altera also makes a (less expensive) version of these devices that is one-time programmable.

10.3.5 Bus Interface Device (PLX464) Family

The last example of a newer type of PLD that can be found is high-drive current types of devices like the PLX Technology PLX464. These devices contain an internal architecture very similar to that of a conventional PAL; however, the outputs are capable of driving much higher output currents. The internal architecture of the PLX464 is shown in Figure 10-10. There are nine dedicated inputs with Schmidt trigger detectors. The hysteresis present in the inputs reduces the possibility of unstable outputs under drive conditions. All of the eight input/output pins are registered and capable of driving either 48 or 64 mA. Four outputs drive 64 mA and four are capable of driving up to 48 mA, eliminating the need for additional tristate bus drivers. The I/O pins can also be tristated by using term-driven output enables on each output. The part also features four output flops being driven by one clock source and four flops driven from a second clock source. This makes the implementation of two independent state machines in a single package possible. This type of device is particularly well suited to bus interface applications. PLX Technology has developed a variety of bus interface modules based on the PLX464 which support the following bus interfaces:

1. VMEbus
   A. Master controller with system arbiter
   B. Slave module
   C. Interrupt generator
   D. Interrupt handler
2. VSB bus
   A. Master module
   B. Slave module
3. MULTIBUS I
   A. Bus arbiter
4. MULTIBUS II
   A. PSB II reply-only agent error generator
   B. LBX II reply agent controller
   C. LBX II reply agent address error generator

10.3 PLD Architecture

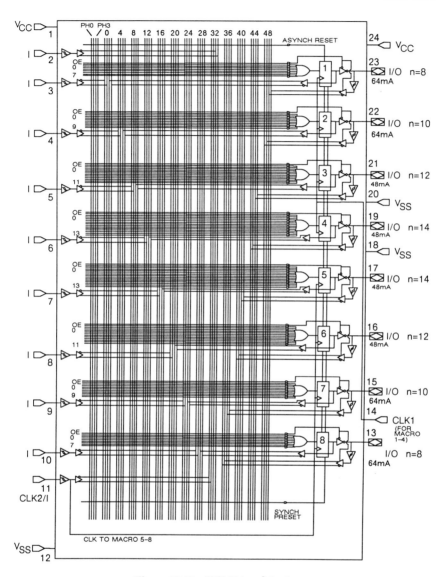

Figure 10-10 PLX 464 architecture.

5. Microchannel bus
      A. Microchannel controller and local arbiter

All of these devices have been preprogrammed by PLX and save the user the necessity of developing and debugging his own program.

10.4 Programming Programmable Logic Devices

There are a variety of techniques that may be used to program a PLD. The simplest devices may be programmed by hand using simple charts and truth tables. The earliest parts were all programmed this way. As the complexity of the parts increased, the need for more sophisticated tools became apparent. One of the earlier tools that allowed programming the parts by simply writing Boolean equations was called PALASM. Copies of this program that run on a PC- or AT-type computer are available at little or no cost from distributors of the logic devices. PALASM and similar programs are, however, somewhat primitive and are analogous (as the name implies) to a hardware assembly language program.

Other tools that are more expensive but also more powerful are also available. An example of this might be ABEL which was developed and is sold by Data I/O Corporation. ABEL supports direct Boolean equation entry but also state machine descriptive entry and provisions for primitive simulation and testing of the parts by means of test vector generation. The simulation is limited to present state/next state or zero delay combinatorial modeling. It does not model physical delays that are present in the part. An example of an ABEL program for a 20RS4 is developed later in this chapter. ABEL also has an advantage in that a great number of different types of devices are supported by the program. The device type is specified in header information at the beginning of the program.

Still other more powerful tools are often required for devices of still greater complexity. Altera Corporation markets a program called APLUS that may be used to develop programs for all of their devices. Unfortunately, it is limited to Altera devices and may not be used for any other manufacturers' parts, unlike ABEL. Like ABEL, it does allow for direct Boolean equation entry but also provides for a higher level design process by use of a feature called LOGIC CAPS in which design may be done using schematic capture and 7400 series logic modules. A design fitter then converts the 7400 device type design into a format that may be programmed into one of the Altera devices. The end product is a single device that "replaces" the parts in the configuration defined by LOGIC CAPS. An example of a design using APLUS is also provided later in this chapter.

The process of defining the contents of a PLD (or LCA) by means of 7400 logic modules has become popular recently. It is a powerful technique and allows rapid and inexpensive development of complex circuits that can ben fit into a single ASIC or PLD. It is my opinion, however, that this process can be easily misused, resulting in sloppy and unreliable designs. Good design practice and discipline should be exercised regardless of implementation technology. State machines and combinatorial functions are best implemented using Boolean equations or state machine entry rather than by a "spaghetti logic" approach to a design. Simulation and timing analysis should be employed whenever possible to ensure that there are no marginal timing parameters.

The final output of nearly every design program is a standardized JEDEC or filename.JED file that nearly all PLD programmers accept as standard input. A PLD programmer accepts this file as the programming specification for determining which fuses are to be blown and left intact in the device or, in the case of EPLDs, which connections are to be made and which are to be left open. Because the format of the JEDEC file is standardized, nearly any PLD programmer that supports the device being programmed may be used with any hardware compiler, such as ABEL or APLUS.

10.5 Metastability

State machine designs that have inputs which may transition at unpredictable intervals have a problem called "metastability." All clocked flip-flops have a metastable window created by the data set-up and hold time. This window is the period of time before and after the rising edge of the clock during which the data input to the flip-flop should not change. This ensures that the output will take on the correct state after the active edge of the clock. If these input conditions are not met and the data transitions during the set-up or hold time, the flip-flop may latch up at either a high or low state for some period of time. Unfortunately, the period of time in which the output is invalid can be far more than one clock period. Several manufacturers have recognized the problem and a number of devices are now characterized for their metastability characteristics.

Xilinx makes a series of field-programmable gate arrays. The internal architecture has a number of flip-flops that currently are specified to support clock rates up to 100 MHz. Data was taken on their XC3020 parts in which a 10-MHz clock was used to load asynchrononous data running at 1 MHz into a flip-flop. The results indicate

Additional output delay	Frequency of occurrence
4.2 ns	Once per hour
6.6 ns	Once per year
8.4 ns	Once per 1000 years

This type of test characterized additional delay only. Other types of failures that can occur are runt pulses, reduced output, slew rate, and oscillation. Devices that exhibit oscillation should be avoided in synchronous machines with asynchronous inputs.

PLX Technology has published a report characterizing the metastability characteristics of a number of different types of flip-flops from discretes and several types of PLDs. The results of their study indicate that the poorest performing parts are the slower ALS, LS, and first-generation programmable logic devices. The faster AS parts and newer GAL devices have better performance, while the best performing parts are their own PLX parts. Their test results are reproduced in Figure 10-11.

In general, for most applications, the issue of metastability is often overstated. There are certainly critical applications where metastability problems can be a very real problem. However, for most applications with currently available fast devices, the probability of a metastability-induced system failure is often quite small. There are ways of (almost) eliminating the problem by using multiphase clocks and gating the input signal with a phase-shifted version of the flip-flop clock. This requires additional parts and logic to implement but improved metastability performance is realized.

10.6 Programming Examples

The following two programming examples illustrate the implementation of a state machine using the Data I/O language called ABEL and Altera Corporation's APLUS. Both implement the state machine whose flow diagram is shown in Figure 10-2. The block diagram of the board showing all inputs and outputs is shown in Figure 10-12. A detailed block diagram of the state machine is shown in Figure 10-13.

The function of the state machine is to handle the interface between a VMEbus and some on-board logic. For this example, a simple analog I/O

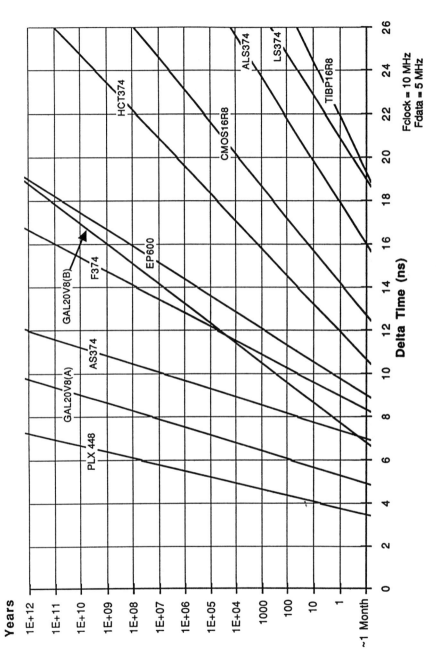

Figure 10-11 MTBF versus Delta.

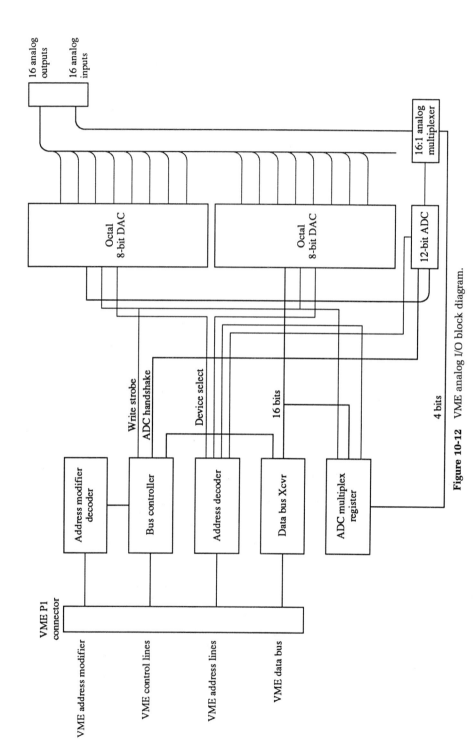

Figure 10-12 VME analog I/O block diagram.

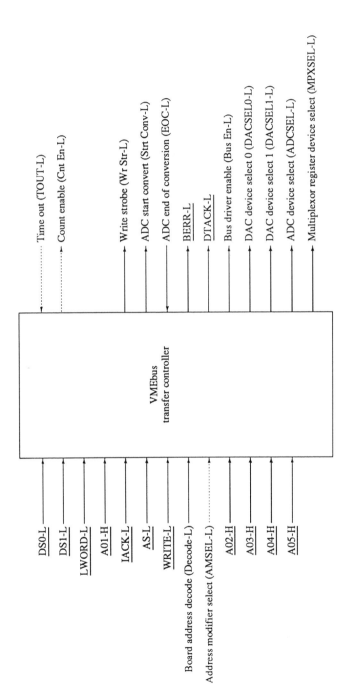

Figure 10-13 VMEbus analog I/O board state machine. Note: (1) All lines shown dashed (- - -) 20RS4 only. (2) Signals underlined are interface to VMEbus. (3) Signals shown heavy (———) EP1800 only. (4) EP1800 design includes VME controller state machine, device select decoding, address modifier decoding, and 4-bit time-out counter (multiplexor stabalization). (5) 20 R54 design includes VME controller state machine.

board was chosen. Such a board might have 8 to 32 digital-to-analog converters (DACs) and one or more analog-to-digital converters (ADCs). Most high-resolution ADCs currently require conversion times from a minimum of a few microseconds to several milliseconds. Devices that require more than 5 to 20 µs to perform a conversion should be handled by means of interrupts. The bus master initiates a conversion process and the ADC slave generates an interrupt when the conversion is complete. For faster converters, the bus master initiates a conversion and the ADC slave performs the conversion by simply reading the ADC. The on-board bus controller initiates the conversion and generates a bus reply when the data become available. The device in this example is assumed to be a "fast" converter. Another feature commonly found is an analog multiplexer driving the input to the ADC. This allows a single ADC to provide conversion for a number of input channels.

The DACs are simpler. Generally, simply writing a value into an internal data register causes the outputs to become stable at the value loaded into the DAC channel. Current technology allows for up to eight 8-bit DACs with registered inputs in one package.

The example VME board contains one 12-bit ADC that converts in less than 5 µs with a 16:1 analog multiplexer feeding the input to the ADC. This provides a 16-channel, 12-bit analog input capability. Additionally, the board has an octal 8-bit DAC. The VME transfer controller state machine contains the logic to interface between the ADC, DACs, and VMEbus. It also provides for on-board control of bus transceivers and on-board "glue" logic. The address map of the board is shown in Figure 10-14. The board has 16 analog input channels and 8 analog output channels in sequential word addresses. The board supports only word transfers. Byte and long-word transfers are excluded here for simplicity. Any attempt to perform long-word or byte transfers will result in bus errors.

There is a two-wire handshake with the ADC. The VME transfer controller generates an active low on StrtConv-L (start convert) to initiate the conversion process simultaneously with a ADCCS-L (ADC chip select) in response to an ADC read command from the bus master. The EOC-L (end of convert) line responds by setting to an inactive high state while the conversion process is in progress. When the ADC is done converting and the output data is ready to be taken, the EOC-L line goes active low. This signals the transfer controller that the ADC data may be transferred to the VMEbus master. The VME transfer controller then places the ADC data on the VMEbus data lines and generates the handshake sequence.

The DAC interface is simpler. The data is presented on the DAC data inputs. In this example, only the low 8 bits of the 16-bit data bus is significant. The DAC is selected by a DACCS-L (DAC chip select) gener-

10.6 Programming Examples

Board address offset (hex word)	Function selected	Device selected	Number of significant bits	Access type	Access size
00	DAC channel	DAC 0	8	R/W	Word (16 bit)
01	DAC channel	DAC 0	8	R/W	Word (16 bit)
02	DAC channel	DAC 0	8	R/W	Word (16 bit)
03	DAC channel	DAC 0	8	R/W	Word (16 bit)
04	DAC channel	DAC 0	8	R/W	Word (16 bit)
05	DAC channel	DAC 0	8	R/W	Word (16 bit)
06	DAC channel	DAC 0	8	R/W	Word (16 bit)
07	DAC channel	DAC 0	8	R/W	Word (16 bit)
08	DAC channel	DAC 1	8	R/W	Word (16 bit)
09	DAC channel	DAC 1	8	R/W	Word (16 bit)
0A	DAC channel	DAC 1	8	R/W	Word (16 bit)
0B	DAC channel	DAC 1	8	R/W	Word (16 bit)
0C	DAC channel	DAC 1	8	R/W	Word (16 bit)
0D	DAC channel	DAC 1	8	R/W	Word (16 bit)
0E	DAC channel	DAC 1	8	R/W	Word (16 bit)
0F	DAC channel	DAC 1	8	R/W	Word (16 bit)
10	A/D converter	ADC	12	RO	Word (16 bit)
11	A/D multiplexor register	ADC Mpx Reg	4	R/W	Word (16 bit)

Figure 10-14 VME analog I/O board address map

ated by the VME transfer controller. Once the data at the input is stable, the transfer controller generates a WrStr-L (write strobe), causing the data to be loaded into the input register for the selected DAC channel. The DAC converts this binary value into a weighted analog output value which is buffered and sent to an output pin on the board.

It can be assumed that the VME controller is a synchronous state machine running at 15 MHz (66.7 ns/clock cycle) rate. The set-up time for the DAC is 100 ns and the hold time is 50 ns. The ADC requires the data to be stable to within 0.5 LSB [least significant bit (0.012%)] before the conversion process is initiated. The multiplexer requires 1.0 μs to stabilize to this accuracy. All ADC channels are read only while the DAC channels are R/W (read/write). The upper eight bits will read as FF (hex). To generate the necessary delays for the DAC set-up time and the ADC multiplexer stabilizing time, a 4-bit binary counter is implemented in the EP1800 in the case of the APLUS example. An external 4-bit counter is used in the ABEL example which implements the state machine with an 20RS4 PLD. See Figure 10-8 for the internal model of the EP1800. The 20RS4 is similar to the 20RS10 shown earlier, except that it has only four registered outputs. Each counter can be enabled and cleared by signals from the state machine. In addition to implementing the time delay counter in the EP1800, all address and address modifier decoding are also done internally. The address map of the prospective board is shown in Figure 10-14.

10.6.1 Programming Example Using APLUS

The APLUS source code has three sections which define special feature, inputs, and outputs (see Figure 10-15). Input and output signals are each preceded by the "INPUTS" and "OUTPUTS" keyword, followed by a list of input and output signals. Note that, unlike ABEL, there need not be a pin assignment performed at this time. APLUS has the ability to perform the assignment. The results of the assignments are given to the report (RPT) file generated by the APLUS design processor. The part is defined in the header and special options may be set to a desired value. The "NETWORK" section of the source code defines characteristics of a given signal. "INP" defines an input signal; "CONF" defines a combinatorial output, no feedback signal; "RORF" defines registered output, registered feedback; "NORF" defines no output registered feedback; and "NOCF" defines no output combinatorial feedback. In the case of the registered signals, the clock source and source of preset or clear inputs is also defined in this section.

The "EQUATIONS" section defines the input conditions required to generate each of the outputs. These are written in standard Boolean form with * indicating "AND" and + indicating inclusive "OR." Each of the four flip-flop inputs to the 4-bit binary counter (CNT0Hd through CNT3Hd) is first defined. The counter has an input called CNTENL. When this signal is low, the counter counts up in normal manner. When high, the counter is held in a cleared state with 0000b at the outputs.

The state machine state assignment is next defined. This assignment is as shown in Figure 10-2. The machine has six states, so three state variables are required to implement the machine. The state variables are called Y2, Y1, and Y0. A few useful intermediate terms, that are used for address decoding and device selection, are then defined. The input conditions required to force a transition from one state to another are then defined as well. These are called the state machine transition conditions. All of these are defined in terms of the current inputs to the machine.

The next part of the "EQUATIONS" section defines the outputs. Some outputs, like the AMSELL signals, are combinatorial outputs of the current inputs. Others such as CNTENL are defined in terms of the current state of the state machine.

10.6 Programming Examples

```
R.M.Cram
Sierra Engineering
Version: 1.0
EP1800J
Analog I/O Board VME Transfer Controller

OPTIONS: TURBO=ON
PART: EP1800

% ++++++++++++++++++++++++++++++++++++++++++++++++++++ %
% VERSION: 1.0            R.M.CRAM                     %
%   INITIAL DEVELOPMENT                                 %
% ++++++++++++++++++++++++++++++++++++++++++++++++++++ %

INPUTS:
        DS0L
        DS1L
        LWORDL
        WRITEL
        IACKL
        ASL
        A01H
        A02H
        A03H
        A04H
        A05H
        AM5H
        AM4H
        AM3H
        AM2H
        AM1H
        AM0H
        EOCL
        DECODEL
        BDCLK

OUTPUTS:
        WRSTRL
        STRTCONVL
        BERRL
        DTACKL
        BUSENL
        DACSEL0L
        DACSEL1L
        ADCSELL
        MPXSELL
        y2,y1,y0                  % STATE MACHINE VARIABLES %

NETWORK:
        AM5H        = INP(AM5H)
        AM4H        = INP(AM4H)
        AM3H        = INP(AM3H)
        AM2H        = INP(AM2H)
        AM1H        = INP(AM1H)
        AM0H        = INP(AM0H)
        A05H        = INP(A05H)
        A04H        = INP(A04H)
        A03H        = INP(A03H)
        A02H        = INP(A02H)
        A01H        = INP(A01H)
```

Figure 10-15 APLUS source code file. *(Figure continues.)*

```
DAC       = /A05H;
DAC0      = /A05H *  /A04H;
DAC1      = /A05H *   A04H;
ADC       =  A05H *  /A04H * /A03H * /A02H * /A01H;
MPX       =  A05H *  /A04H * /A03H * /A02H *  A01H;
DEVSEL    = /DECODEL * /AMSELL * /ASL * /DS0L * /DS1L * LWORDL;

% +++++++++++++++++++++++++++++++++++++++++++++++++++++ %
% STATE MACHINE TRANSITION CONDITIONS                   %
% +++++++++++++++++++++++++++++++++++++++++++++++++++++ %

MPXWR    = MPX * DEVSEL * /WRITEL * IACKL;
ADCREAD  = ADC * DEVSEL *  WRITEL * IACKL;
DACTR    = DAC * DEVSEL *  IACKL;
MPXRD    = MPX * DEVSEL *  WRITEL * IACKL;
ADCBUSY  = ADC * DEVSEL *  WRITEL * EOCL;
READ     = WRITEL;
DACWR    = DAC * DEVSEL * /WRITEL;
VMEDONE  = DS0L * DS1L;
TOUT     = CNT0H * CNT1H * CNT2H * CNT3H;

% +++++++++++++++++++++++++++++++++++++++++++++++++++++ %
% OUTPUT EQUATIONS                                      %
% +++++++++++++++++++++++++++++++++++++++++++++++++++++ %

/AMSELLc    = (  AM5H *  AM4H *  AM3H *  AM2H * /AM1H *  AM0H )
            + (  AM5H * /AM4H *  AM3H * /AM2H * /AM1H *  AM0H )
            + ( /AM5H *  AM4H *  AM3H *  AM2H * /AM1H *  AM0H )
            + ( /AM5H * /AM4H *  AM3H *  AM2H * /AM1H *  AM0H );

/CNTENLc    = TIMEOUT;

/WRSTRLc    = WRSTROBE;

/STRTCONVLc = STRTADC;

/BERRLc     = VCC;

/DTACKLc    = HANDSHAKE;

BUSENLc     = IDLE;

/DACSEL0Lc  = DAC0 * DEVSEL;

/DACSEL1Lc  = DAC1 * DEVSEL;

/ADCSELLc   = ADC * DEVSEL;

/MPXSELLc   = MPX * DEVSEL;

% +++++++++++++++++++++++++++++++++++++++++++++++++++++ %
% STATE VARIABLE EQUATIONS                              %
% +++++++++++++++++++++++++++++++++++++++++++++++++++++ %

Y2d = (SELECT * (DACWR + MPXWR + READ)) + WRSTROBE + (HANDSHAKE * /VMEDONE)

Y1d = (IDLE * (ADCREAD + DACTR + MPXRD)) + STRTADC + (SELECT * ADCBUSY)
    + (SELECT * (DACWR + MPXWR)) + (TIMEOUT * TOUT);
Y0d = (IDLE * (MPXWR + DACTR + MPXRD)) + TIMEOUT + (SELECT * ADCBUSY)
    + (SELECT * (DACWR + MPXWR));

END$
```

Figure 10-15 *(continued)*

The last part of this section is the state variable equations. These three equations are the heart of the state machine. They define the movement of the machine from the current state to the next state and the conditions required for this transition. Unlike the output equations, these equations are written as transition equations. Each transition may be thought of as a sum of products in which a transition may be forced by the occurrence of a given current state concurrent with a specified input condition. The Y2 equation may be read as follows:

Set the Y2 variable to high if

1. The current state is SELECT and the input is DACWR or MPXWR or READ

   OR
2. The current state is WRSTROBE

   OR
3. The current state is HANDSHAKE and the input is not VMEDONE

Once you are familiar with the notation, the equations read as English statements.

Once the source file is generated using a text editor, the APLUS ADP (Altera Design Processor) is run on the source code. Assuming no mistakes have been made (usually a wrong assumption), two output files are generated. The JED (JEDEC) file is the hex bit map file used to program the physical device. The RPT (REPORT) file contains information on the utilization and processing of the design file. A copy of the RPT file for this design has been reproduced here (Figure 10-16). The first section contains information on options enabled and header data included at the beginning of the source code. Next a pictorial view of the part is produced, showing which signals were assigned to which pins. This is used to design the printed circuit board on which the part is to be installed. The next section contains information on register usage and types and where the register is physically located in the device. Inputs are also defined and unused resources are listed. Part utilization is also produced. In the case of this design, 37% of the macrocells, 90% of the input pins, and 29% of the product terms were used. The last section contains interconnection cross-reference information. This information is useful for debugging designs and displaying data on the internal interconnection resources used in the device.

10.6.2 Programming Example Using ABEL

ABEL, which is sold and supported by Data I/O Corporation, has the advantage of supporting a wide spectrum of devices. ABEL also supports

```
            ALTERA Design Processor Utilization Report                         VME.rpt
            @(#) FIT Version 5.02    11/10/87 20:16:31 39.16
            ***** Design implemented successfully

            R.M.Cram
            Sierra Engineering
            Version: 1.0
            EP1800J
            Analog I/O Board VME Transfer Controller

            Input files : VME.adf
            ADP Options: Minimization = Yes,   Inversion Control = No,   LEF Analysis = No

            OPTIONS: TURBO = ON, SECURITY = OFF
               ***** Externally connect signal "BDCLK" to pins 17 and 53.

                                                       S
                                D  D                   T
                                A  A                   R
                             M  C  C              A    T
                             P  S  S        B     D    W  T  D  B
                             X  E  E        E     C    R  C  T  U
                             S  L  L  G  G  R  G  E  G T  O  A  S
                             E  1  0  n  n  R  n  L  N R  V  K  N  G  G  G  G
                             L  L  L  d  d  L  d  L  D L  L  L  L  n  n  n  n
                             L                                       d  d  d  d
                             ---------------------------------------------------
                            / 9  8  7  6  5  4  3  2  1 68 67 66 65 64 63 62 61 |
                    ASL  | 10                                                60 | RESERVED
                    y0   | 11                                                59 | RESERVED
                    y1   | 12                                                58 | AM5H
                    y2   | 13                                                57 | AM4H
                    Gnd  | 14                                                56 | WRITEL
                    Gnd  | 15                                                55 | LWORDL
                    A01H | 16                                                54 | IACKL
                    BDCLK| 17                                                53 | BDCLK
                    Vcc  | 18                                                52 | Vcc
                    A02H | 19                                                51 | EOCL
                    A03H | 20                                                50 | DS1L
                    A04H | 21                                                49 | DS0L
                    A05H | 22                                                48 | DECODEL
                    AM0H | 23                                                47 | Gnd
                    AM1H | 24                                                46 | Gnd
                    AM2H | 25                                                45 | Gnd
                    AM3H | 26                                                44 | Gnd
                         |_ 27 28 29 30 31 32 33 34 35 36 37 38 39 40 41 42 43 _|
                             ---------------------------------------------------
                             G  G  G  G  G  G  G  G  G  G  G  G  G  G  G  G  G
                             n  n  n  n  n  n  n  n  N  n  n  n  n  n  n  n  n
                             d  d  d  d  d  d  d  d  D  d  d  d  d  d  d  d  d

                                                                               VME.rpt
            **OUTPUTS**
                                                        FdBck
                  Name    Pin  Resource  MCell  PTerms | Group | Sync Clock | OE Group

                ADCSELL    2   CONF       1     1/ 8  |   1   |      -     |  VCC
                BERRL      4   CONF       3     1/ 8  |   1   |      -     |  VCC
                BUSENL    65   CONF      45     1/ 8  |   4   |      -     |  VCC
                DACSEL0L   7   CONF       6     1/ 8  |   1   |      -     |  VCC
                DACSEL1L   8   CONF       7     1/ 8  |   1   |      -     |  VCC
                DTACKL    66   CONF      46     1/ 8  |   4   |      -     |  VCC
                MPXSELL    9   CONF       8     1/ 8  |   1   |      -     |  VCC
                STRTCONVL 67   CONF      47     1/ 8  |   4   |      -     |  VCC
                WRSTRL    68   CONF      48     1/ 8  |   4   |      -     |  VCC
                y0        11   ROIF!     10     6/ 8  |  1G   |   BDCLK    |  VCC
                y1        12   ROIF!     11     7/ 8  |  1G   |   BDCLK    |  VCC
                y2        13   ROIF!     12     6/ 8  |  1G   |   BDCLK    |  VCC


            **BURIED REGISTERS**
                                                        FdBck
                  Name    Pin  Resource  MCell  PTerms | Group | Sync Clock | OE Group

                AMSELL  ( 3)   NOCF       2     3/ 8  |   1   |      -     |    -
                CNTENL  (59)   COIF!     39     1/ 8  |  4G   |      -     |  VCC
                CNT0H   (60)   ROIF!     40     1/ 8  |  4G   |   BDCLK    |  VCC
                CNT1H   ( 5)   NORF       4     2/ 8  |   1   |   BDCLK    |    -
                CNT2H   ( 6)   NORF       5     3/ 8  |   1   |   BDCLK    |    -
                CNT3H   (10)   NORF       9     4/ 8  |   1   |   BDCLK    |    -
```

Figure 10-16   APLUS report file. *(Figure continues.)*

10.6 Programming Examples

```
**INPUTS**
                                      FdBck
    Name    Pin  Resource  MCell  PTerms  Group  Sync Clock  OE Group
    AM0H    23   INP         -      -       -        -          -
    AM1H    24   INP         -      -       -        -          -
    AM2H    25   INP         -      -       -        -          -
    AM3H    26   INP         -      -       -        -          -
    AM4H    57   INP         -      -       -        -          -
    AM5H    58   INP         -      -       -        -          -
     ASL    10   INP         -      -       -        -          -
    A01H    16   INP         -      -       -        -          -
    A02H    19   INP         -      -       -        -          -
    A03H    20   INP         -      -       -        -          -
    A04H    21   INP         -      -       -        -          -
    A05H    22   INP         -      -       -        -          -
   BDCLK    53   CKR         -      -       4        -          -
 DECODEL    48   INP         -      -       -        -          -
    DS0L    49   INP         -      -       -        -          -
    DS1L    50   INP         -      -       -        -          -
    EOCL    51   INP         -      -       -        -          -
   IACKL    54   INP         -      -       -        -          -
  LWORDL    55   INP         -      -       -        -          -
  WRITEL    56   INP         -      -       -        -          -

**UNUSED RESOURCES**
                                      FdBck
    Name    Pin  Resource  MCell  PTerms  Group  Sync Clock  OE Group
     -      14   INPUT       -      -       -        -          -
     -      15   INPUT       -      -       -        -          -
     -       -   MCELL      13      8       2        -          -
     -       -   MCELL      14      8       2        -          -
     -       -   MCELL      15      8       2        -          -
     -       -   MCELL      16      8       2        -          -
     -      27   MCELL      17      8       2        -          -
     -      28   MCELL      18      8       2        -          -
     -      29   MCELL      19      8       2        -          -
     -      30   MCELL      20      8       2        -          -
     -      31   MCELL      21      8       2        -          -
     -      32   MCELL      22      8       2        -          -
     -      33   MCELL      23      8       2        -          -
     -      34   MCELL      24      8       2        -          -
     -      36   MCELL      25      8       3        -          -
     -      37   MCELL      26      8       3        -          -
     -      38   MCELL      27      8       3        -          -
     -      39   MCELL      28      8       3        -          -
     -      40   MCELL      29      8       3        -          -
     -      41   MCELL      30      8       3        -          -
     -      42   MCELL      31      8       3        -          -
     -      43   MCELL      32      8       3        -          -
     -      44   MCELL      33      8      3G        -          -
     -      45   MCELL      34      8      3G        -          -
     -      46   MCELL      35      8      3G        -          -
     -      47   MCELL      36      8      3G        -          -
     -       -   MCELL      37      8       4      BDCLK        -
     -       -   MCELL      38      8       4      BDCLK        -
     -      61   MCELL      41      8       4      BDCLK        -
     -      62   MCELL      42      8       4      BDCLK        -
     -      63   MCELL      43      8       4      BDCLK        -
     -      64   MCELL      44      8       4      BDCLK        -

**PART UTILIZATION**

18/48  MacroCells (37%)
19/21  Input Pins (90%)
       PTerms Used 29%
```

Figure 10-16 (*continued*)

```
Macrocell Interconnection Cross Reference

FEEDBACKS:                              M M M   M M M M M M
                M M M M M M M M M 1 1 1 3 4 4 4 4 4
                1 2 3 4 5 6 7 8 9 0 1 2 9 0 5 6 7 8
ADCSELL .. CONF @M1 ->  . . . . . . . . . . . .   x x x x x x @2
AMSELL ... NOCF @M2 ->  * . . . . * * * . * * *   x x x x x x (3)
BERRL .... CONF @M3 ->  . . . . . . . . . . . .   x x x x x x @4
CNT1H .... NORF @M4 ->  . . . * * . . . . * . *   x x x x x x (5)
CNT2H .... NORF @M5 ->  . . . . * . . . . * . *   x x x x x x (6)
DACSEL0L . CONF @M6 ->  . . . . . . . . . . . .   x x x x x x @7
DACSEL1L . CONF @M7 ->  . . . . . . . . . . . .   x x x x x x @8
MPXSELL .. CONF @M8 ->  . . . . . . . . . . . .   x x x x x x @9
CNT3H .... NORF @M9 ->  . . . . . * . . . * . *   x x x x x x (10)
y0 ....... ROIF @M10->  . . . . . . . . . * * *   * . * * * * @11
y1 ....... ROIF @M11->  . . . . . . . . . * * *   * . * * * * @12
y2 ....... ROIF @M12->  . . . . . . . . . * * *   * . * * * * @13

CNTENL ... COIF @M39->  . . . * * . . . . * . .   . * . . . . (59)
CNT0H .... ROIF @M40->  . . . * * . . . . * . .   . * . . . . (60)
BUSENL ... CONF @M45->  x x x x x x x x x x x x   . . . . . . @65
DTACKL ... CONF @M46->  x x x x x x x x x x x x   . . . . . . @66
STRTCONVL  CONF @M47->  x x x x x x x x x x x x   . . . . . . @67
WRSTRL ... CONF @M48->  x x x x x x x x x x x x   . . . . . . @68
                        A A B C C D D M C y y y   C C B D S W
                        D M E N N A A P N 0 1 2   N N U T T R
                        C S R T T C C X T         T T S A R S
                        S E R 1 2 S S S 3         E 0 E C T T
                        E L L H H E E E H         N H N K C R
                        L L     L L L             L   L L O L
                        L       0 1 L                       N
                                L   L                       V
                                                            L

Macrocell Interconnection Cross Reference

INPUTS:                                 M M M   M M M M M M
                M M M M M M M M M 1 1 1 3 4 4 4 4 4
                1 2 3 4 5 6 7 8 9 0 1 2 9 0 5 6 7 8
ASL ...... INP  @10 ->  * . . . . * * . * * * *   . . . . . . M9
A01H ..... INP  @16 ->  * . . . . . . * . * * *   . . . . . .
A02H ..... INP  @19 ->  * . . . . . . * . * * *   . . . . . .
A03H ..... INP  @20 ->  * . . . . . . * . * * *   . . . . . .
A04H ..... INP  @21 ->  * . . . . * * . . * * *   . . . . . .
A05H ..... INP  @22 ->  * . . . . * * * . * * *   . . . . . .
AM0H ..... INP  @23 ->  . * . . . . . . . . . .   . . . . . . M13
AM1H ..... INP  @24 ->  . * . . . . . . . . . .   . . . . . . M14
AM2H ..... INP  @25 ->  . * . . . . . . . . . .   . . . . . . M15
AM3H ..... INP  @26 ->  . * . . . . . . . . . .   . . . . . . M16
DECODEL .. INP  @48 ->  * . . . . * * * . * * *   . . . . . .
DS0L ..... INP  @49 ->  * . . . . * * * . * * *   . . . . . .
DS1L ..... INP  @50 ->  * . . . . * * * . * * *   . . . . . .
EOCL ..... INP  @51 ->  . . . . . . . . . * * .   . . . . . .
IACKL .... INP  @54 ->  . . . . . . . . . * * *   . . . . . .
LWORDL ... INP  @55 ->  * . . . . * * * . * * *   . . . . . .
WRITEL ... INP  @56 ->  . . . . . . . . . * * *   . . . . . .
AM4H ..... INP  @57 ->  . * . . . . . . . . . .   . . . . . . M37
AM5H ..... INP  @58 ->  . * . . . . . . . . . .   . . . . . . M38
                        A A B C C D D M C y y y   C C B D S W
                        D M E N N A A P N 0 1 2   N N U T T R
                        C S R T T C C X T         T T S A R S
                        S E R 1 2 S S S 3         E 0 E C T T
                        E L L H H E E E H         N H N K C R
                        L L     L L L             L   L L O L
                        L       0 1 L                       N
                                L   L                       V
                                                            L
```

Figure 10-16 *(continued)*

10.6 Programming Examples

a variety of input specifications, including Boolean equations, state machine descriptions, and truth tables.

The design example shown here (Figure 10-17) implements the same design as the APLUS example shown previously; however, the design is implemented in a 20RS4. Since the device is considerably smaller, only the VME state machine has been implemented. It is assumed that the address decoder, address modifier decoder, and counter have been implemented in other logic. The first section of the source code defines the module name, title, and device type.

ABEL does not have the ability to define pins and as such all signals must be assigned to a pin. The next section defines all input and output signal names and assigns pin numbers to the signal names. The 20RS4 has four registered outputs, of which three are used for state variables. One is used as a signal output. The three state variables are Y2, Y1, Y0 and the registered output is the write strobe (WRSTRL).

One of the more useful features supported by ABEL is the ability to define signal sets. This feature is used to identify the machine state assignments or MSTATE. The state name IDLE is assigned the value Y2 = 0, Y1 = 0, and Y0 = 0. In similar manner, the state name SELECT has been assigned the value Y2 = 0, Y1 = 1, and Y0 = 1.

In a manner similar to the assignment of state variables, the transition conditions (TC) have been defined in terms of all of the state machine inputs. An "X" in a position is a don't care condition. For example, the ADCREAD transition condition becomes valid when the following inputs are

$$
\begin{aligned}
DS1L &= 0 \\
DS0L &= 0 \\
LWORDL &= 1 \\
WRITEL &= 1 \\
DECODEL &= 0 \\
AMSELL &= 0 \\
IACKL &= 1 \\
ASL &= 0 \\
A05H &= 1 \\
A01H &= 0 \\
EOCL &= \text{DON'T CARE} \\
TOUTL &= \text{DON'T CARE}
\end{aligned}
$$

The last section defines the Boolean equations that generate the outputs, the "=" symbol indicates that the output is combinatorial, while the ":=" symbol indicates that the output is registered. The "&" symbol is an AND and the "#" symbol is inclusive OR. Note that the outputs are

```
module VME
title 'VME Analog I/O Board Bus Controller'

        VMECON    device    'P20RS4';

" +++++++++++++++++++++++++++++++++++++++++++++++++++++++++++
" This source code and ABEL output reproduced courtesy and
" with permission of DATA I/O Corporation.
" +++++++++++++++++++++++++++++++++++++++++++++++++++++++++++

" +++++++++++++++++++++++++++++++++++++++++++++++++++++++++++
" Version:      1.0              R.M.Cram
"   Initial development - ABEL design example
" +++++++++++++++++++++++++++++++++++++++++++++++++++++++++++

Ck,X = .C.,.X.;

" +++++++++++++++++++++++++++++++++++++++++++++++++++++++++++
" Control Inputs:
" +++++++++++++++++++++++++++++++++++++++++++++++++++++++++++

        CLK             pin     1;

" +++++++++++++++++++++++++++++++++++++++++++++++++++++++++++
" VME Inputs:
" +++++++++++++++++++++++++++++++++++++++++++++++++++++++++++

        WRITEL          pin     2;
        DS0L            pin     3;
        DS1L            pin     4;
        LWORDL          pin     5;
        IACKL           pin     6;
        DECODEL         pin     7;
        AMSELL          pin     8;
        A01H            pin     9;
        A05H            pin     10;
        ASL             pin     11;

" +++++++++++++++++++++++++++++++++++++++++++++++++++++++++++
" ADC Input:
" +++++++++++++++++++++++++++++++++++++++++++++++++++++++++++

        EOCL            pin     14;

" +++++++++++++++++++++++++++++++++++++++++++++++++++++++++++
" Timer Input:
" +++++++++++++++++++++++++++++++++++++++++++++++++++++++++++

        TOUTL           pin     23;

" +++++++++++++++++++++++++++++++++++++++++++++++++++++++++++
" Outputs:
" +++++++++++++++++++++++++++++++++++++++++++++++++++++++++++

        BUSENL          pin     22;
        DTACKL          pin     21;
        y2              pin     20;
        y1              pin     19;
        y0              pin     18;
        WRSTRL          pin     17;
```

Figure 10-17 ABEL source code file. *(Figure continues.)*

10.6 Programming Examples

```
                STRTCONVL       pin     16;
                CNTENL          pin     15;

" +++++++++++++++++++++++++++++++++++++++++++++++++++++++++
" Machine State Assignment:
" +++++++++++++++++++++++++++++++++++++++++++++++++++++++++

                MSTATE          = [y2,y1,y0];

                IDLE            = [0,0,0];
                TIMEOUT         = [0,0,1];
                STRTADC         = [0,1,0];
                SELECT          = [0,1,1];
                HANDSHAKE       = [1,0,0];
                WRSTROBE        = [1,1,1];

" +++++++++++++++++++++++++++++++++++++++++++++++++++++++++
" Transition Conditions:
" +++++++++++++++++++++++++++++++++++++++++++++++++++++++++

TC = [DS1L,DS0L,LWORDL,WRITEL,DECODEL,AMSELL,IACKL,ASL,A05H,A01H,EOCL,TOUTL];

                MPXWR    = [0,0,1,0,0,0,1,0,1,1,X,X];
                DACTR    = [0,0,1,X,0,0,1,0,0,X,X,X];
                MPXRD    = [0,0,1,1,0,0,1,0,1,1,X,X];
                ADCREAD  = [0,0,1,1,0,0,1,0,1,0,X,X];
                TOUT     = [X,X,X,X,X,X,X,X,X,X,X,0];
                ADCBUSY  = [X,X,X,X,X,X,X,X,X,X,1,X];
                READ     = [X,X,1,X,X,X,X,X,X,X,X,X];
                VMEDONE  = [1,1,1,X,X,X,X,X,X,X,X,X];
                DACWR    = [0,0,1,0,0,0,X,X,X,X,X,X];

Equations

" +++++++++++++++++++++++++++++++++++++++++++++++++++++++++
" Output equations:
" +++++++++++++++++++++++++++++++++++++++++++++++++++++++++

                BUSENL    = (MSTATE == IDLE);

                !DTACKL   = (MSTATE == HANDSHAKE);

                !WRSTRL  := ((MSTATE == SELECT) & (TC == (DACWR # MPXWR)));

                !STRTCONVL= (MSTATE == STRTADC);

                !CNTENL   = (MSTATE == TIMEOUT);

" +++++++++++++++++++++++++++++++++++++++++++++++++++++++++
" State variable equations:
" +++++++++++++++++++++++++++++++++++++++++++++++++++++++++

            y2 := ((MSTATE == SELECT) & (TC == (DACWR # MPXWR # READ)))
               # (MSTATE == WRSTROBE)
               # ((MSTATE == HANDSHAKE) & (TC == !VMEDONE));

            y1 := ((MSTATE == IDLE) & (TC == (ADCREAD # DACTR # MPXRD)))
               # (MSTATE == STRTADC)
               # ((MSTATE == TIMEOUT) & (TC == TOUT))
               # ((MSTATE == SELECT) & (TC == (DACWR # MPXWR # ADCBUSY)));

            y0 := ((MSTATE == IDLE) & (TC == MPXWR))
               # ((MSTATE == STRTADC) & (TC == ADCBUSY))
               # ((MSTATE == SELECT) & (TC == (ADCBUSY # DACWR # MPXWR)));

end VME
```

Figure 10-17 *(continued)*

written in terms of the current state of the machine, with the exception of the WRSTRL signal which is written in terms of transition conditions. Since this is a registered output, the output would be delayed by one clock cycle if the equation were in terms of current state. Once the inputs become valid, it takes one clock cycle for the machine to transition to the correct next state. Registered outputs require one additional clock cycle to become valid after the machine has entered the new state. This additional clock cycle delay can be avoided if the output is written in terms of the old state and the input transition condition required to move to the new state. This typically requires more product terms but in the PLD this is not important unless the number of product terms exceeds the capacity of the output OR gate. Other than the syntax of the statements, they can be read and are implemented very similarly to the APLUS example provided previously.

After the source code file has been implemented, a variety of post-processing programs are executed on the source code, generating a num-

```
                                                               Page 1
ABEL(tm) Version 2.00b  -  Document Generator       02-Dec-89 02:10 PM
VME Analog I/O Board Bus Controller
Equations for Module VME

Device VMECON

    Reduced Equations:

    BUSENL = (!y0 & !y1 & !y2);

    DTACKL = !(!y0 & !y1 & y2);

    WRSTRL := !(A01H & A05H & !AMSELL & !ASL & !DECODEL & !DSOL &
               !DS1L & IACKL & LWORDL & !WRITEL & y0 & y1 & !y2);

    STRTCONVL = !(!y0 & y1 & !y2);

    CNTENL = !(y0 & !y1 & !y2);

    y2 := (y0 & y1 & y2
         # !DSOL & !DS1L & !LWORDL & !y0 & !y1 & y2
         # A01H & A05H & !AMSELL & !ASL & !DECODEL & !DSOL & !DS1L &
           IACKL & LWORDL & !WRITEL & y0 & y1 & !y2);

    y1 := (!y0 & y1 & !y2
         # !TOUTL & y0 & !y1 & !y2
         # A01H & A05H & !AMSELL & !ASL & !DECODEL & !DSOL & !DS1L &
           IACKL & LWORDL & WRITEL & !y0 & !y1 & !y2
         # A01H & A05H & !AMSELL & !ASL & !DECODEL & !DSOL & !DS1L &
           EOCL & IACKL & LWORDL & !WRITEL & y0 & y1 & !y2);

    y0 := (EOCL & !y0 & y1 & !y2
         # A01H & A05H & !AMSELL & !ASL & !DECODEL & !DSOL & !DS1L &
           IACKL & LWORDL & !WRITEL & !y0 & !y1 & !y2
         # A01H & A05H & !AMSELL & !ASL & !DECODEL & !DSOL & !DS1L &
           EOCL & IACKL & LWORDL & !WRITEL & y0 & y1 & !y2);
```

Figure 10-18 ABEL document file. *(Figure continues.)*

10.6 Programming Examples

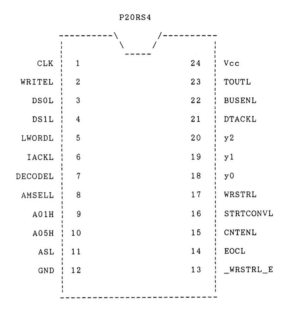

```
ABEL(tm) Version 2.00b  -  Document Generator
VME Analog I/O Board Bus Controller
Chip diagram for Module VME

Device VMECON

                         P20RS4
                ---------\   /----------
                          \_/
                          ---
              CLK  |  1            24  | Vcc
           WRITEL  |  2            23  | TOUTL
             DSOL  |  3            22  | BUSENL
             DS1L  |  4            21  | DTACKL
           LWORDL  |  5            20  | y2
            IACKL  |  6            19  | y1
          DECODEL  |  7            18  | y0
           AMSELL  |  8            17  | WRSTRL
             A01H  |  9            16  | STRTCONVL
             A05H  | 10            15  | CNTENL
              ASL  | 11            14  | EOCL
              GND  | 12            13  | _WRSTRL_E
                --------------------------

end of module VME
```

Figure 10-18 *(continued)*

ber of different output files. Most of these are not of interest unless difficulties are encountered in the reduction process. As with APLUS, the bit map file is the .JED file and is required to program the device. Note also that the 20RS4 is a fuse device and can be programmed only once. ABEL supports simulation features that are not demonstrated here. These simulation features are very convenient and can save both parts and debugging time. One convenient file created is the document or .DOC file. The document file for this design is reproduced in Figure 10-18. It contains the reduced form of the equations that are actually implemented in the fuse map and directly correlate to blown and unblown fuses. A picture of the device, along with pin assignments, is also produced as a part of the document file.

Field-Programmable Gate Arrays

11.0 Field-Programmable Gate Arrays Overview

This chapter is organized a little differently from the previous chapters, due to a limitation in space. A few specific devices will be presented to highlight their internal architecture and capabilities. It is hoped that by exposing the reader to a (very) few specific device examples, a generalized understanding of the capabilities of FPGA design will be realized. The FPGA market is maturing and evolving rapidly and the devices used here may well be replaced by faster, more powerful, and lower cost devices within a matter of months. However, the basic principles behind gate array design will remain stable for many years to come and are independent of device type. By examining specific devices, a general understanding may be reached.

Field-programmable gate arrays, or FPGAs, are an important step up in integration density from the PLDs. FPGAs do not replace PLDs or make them obsolete, but they are an additional powerful tool the designer has in a spectrum of "screwdrivers" that are used to implement a design solution. In general, FPGAs have an internal architecture that is more register-oriented than PLDs and are more poorly suited for random logic or state machine types of problems than PLDs. They do contain a greater number of equivalent gates at the upper end of device complexity. Currently, densities can range up to about 15,000 to 20,000 gates for high-end FPGAs while PLDs are currently at about the 5000- to 8000-gate level. Gate count can be misleading, however. If a design is register-intensive, an FPGA should certainly be considered but if it is random-logic-intensive, a PLD may be a better selection.

Cost sensitivity and anticipated production volume play a major role in device selection also. While PLDs may be more cost effective and the development tools lower in price for limited production run items, an FPGA can be quite expensive in small volumes but drops rapidly in price as volume increases. Since gate count tends to be higher in the FPGAs, a greater number of functions may be "stuffed" into a single device, reducing the amount of board space required to implement a given function. In general, the FPGA market is maturing rapidly and prices are falling rapidly. Devices are becoming more powerful and design, simulation, and debugging tools more sophisticated. The parts are becoming increasingly more attractive even for small production runs under the right design constraints.

11.1 FPGA Speed Considerations

The speed with which an FPGA operates is more difficult to specify than for a PLD. It is highly dependent upon internal architecture, routing complexity, and device capability. For example, Xilinx, a maker of FPGAs, specifies the speed of operation by the maximum toggle frequency of any internal flip-flop. Currently, their toggle rates range up to 100 MHz and can be expected to increase rapidly in the next few years. When the device is internally configured into an architecture that implements the desired function, the speed of the silicon "system" will decrease. An input to valid output signal is routed through an input pad, internal "metalization" layer to a logic function, propagates through the intended logic function that may exist in several layers, is routed to an output pad, and appears on the output pin of the device. Each stage requires some propagation time. The total system speed is derived from the sum of the delays through each element of the system. Typical speeds of real-world designs for a 100-MHz toggle rate device would permit operation in the 25- to 35-MHz range. Most development software has provisions for predicting device speeds after the design has been implemented. This is part of the simulation function that is highly recommended for an FPGA design of any complexity.

11.2 FPGA Architecture

Individual devices have different internal architectural details; however, as in PLDs, there is a lot of similarity between devices. There are a variety

11.2 FPGA Architecture

of companies making FPGAs, including Advanced Micro Devices (AMD), ACTEL, and Xilinx. Xilinx currently markets two series of devices: the XC2000 series, which was the first to be commercially produced, and the XC3000 series, with plans to release the XC4000 very soon. The parts can be characterized at four levels:

1. Logic capacity in usable gates
2. Number of and capability of internal logic blocks
3. Number of user I/O pins
4. Interconnect architecture and power

A weakness in any one of these four levels can produce a part that will not be capable of meeting a given design requirement. The Xilinx parts in these two families have a usable gate capacity ranging from 1200 to 9000 gates. Internally, all parts consist of an array of configurable logic blocks, or CLBs. The CLB can be thought of as a miniature PLD in that it contains flip-flops, a combinatorial function module implemented using a PROM technique, and several multiplexers. The number of CLBs ranges from 64 to 320 per device. Signals are brought into and taken out of the device by means of user I/O pads. All inputs and outputs must be buffered.

11.2.1 I/O Characteristics

Nearly all I/O pads are universal pads that may be configured for any of the following options:

1. Combinatorial or direct input
2. Edge-triggered flip-flop
3. Level-sensitive latch
4. CMOS or TTL threshold
5. Combinatorial or direct output
6. Registered output
7. Inverted output
8. Tristate output
9. Slew rate limited output

The architecture of the Xilinx XC3000 series family CLB is shown in Figure 11-1. The architecture of the programmable I/O block is shown in Figure 11-2 for the same family. These represent an increase in capability (and complexity) from the CLB and I/O block for the XC2000 family of parts shown in Figures 11-3 and 11-4.

The number of available user I/O pins ranges from a low of 58 for the smallest XC2000 family member (the XC2064) to a high of 144 pins for the largest XC3000 family member (the XC3090). The I/O blocks and pins or

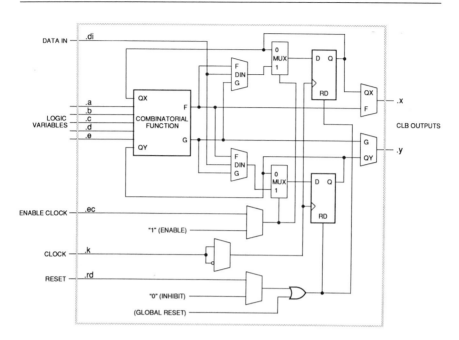

Figure 11-1 Xilinx XC3000 series configurable logic block. Each configurable logic block includes a combinatorial logic section, two flip-flops, and a program memory–controlled multiplexer selection of function. It has five logic variable inputs .a, .b, .c, .d, and .e; a direct data in .di; an enable clock .ec; a clock (invertible) .k; an asynchronous reset .rd; and two outputs .x and .y.

pads are located physically around the periphery of the device. The interior of the device is an array of CLB elements. Each CLB element can implement limited logic functions.

11.2.2 Interconnection Resources

By interconnecting multiple CLBs, more complex high-level functions can be implemented. Those high-level functions are then connected into even higher level functions, continuing on in this manner until the final highest level of functionality is achieved. This type of design methodology is a "hierarchical" approach to design. It is a useful and almost required method for gate array design. It breaks the design into manageable levels and allows different designers to work independently before combining the end result. It is even possible to breadboard subsets of the final design in silicon and test those elements before integrating all parts of the design into a final product.

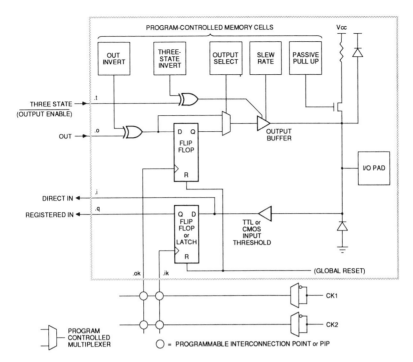

Figure 11-2 Xilinx XC3000 series input/output block. The input/output block includes input and output storage elements and I/O options selected by configuration memory cells. A choice of two clocks is available on each die edge. The polarity of each clock line (not each flip-flop or latch) is progammable. A clock line that triggers the flip-flop on the rising edge is an active low latch enable (latch transparent) signal and vice versa. Passive pull-up can only be enabled on inputs, not on outputs. All user inputs are programmed for TTL or CMOS thresholds.

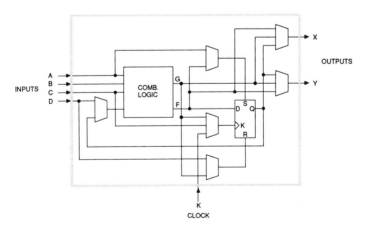

Figure 11-3 Xilinx XC2000 series configurable logic block.

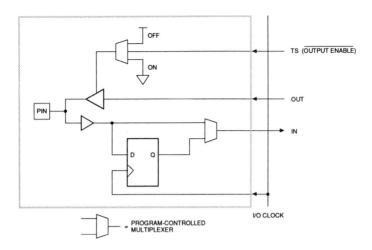

Figure 11-4 Xilinx XC2000 series input/output block.

The internal interconnection capability of the Xilinx XC3000 family exists on three levels with a pseudo fourth level. The first type of interconnection is called general-purpose interconnect and consists of a grid of segments that are interconnected through switch matrices. These matrices are located between each of the CLB elements, as shown in Figure 11-5. The general-purpose interconnect lines can be buffered by bidirectional buffers. The general-purpose interconnect lines can be used to connect any CLB or I/O block to any other CLB or I/O block. The delays through the buffers and switch matrices may be objectionable for elements that are widely spaced on the device. Each matrix has five vertical and five horizontal bidirectional I/O lines at each matrix node.

The second level is the direct interconnect level. Signals routed using direct interconnect have the smallest propagation delay and do not use general interconnect lines. Each CLB's .x output can be routed to the .b input of the CLB to its right, and the .c input of the CLB directly to its left. The .y output can connect directly to the .d input of the CLB directly above and the .a input of the CLB directly below. This interconnection between CLBs is the most desirable to use in that it results in fastest performance; however, it is limited to connecting only blocks that are the next neighbors of the output block. These lines may not be tristated; only selected inputs may be driven from the output CLB. This type of interconnect is shown in Figure 11-6. See Figure 11-1 for identification of input and output definitions.

11.2 FPGA Architecture

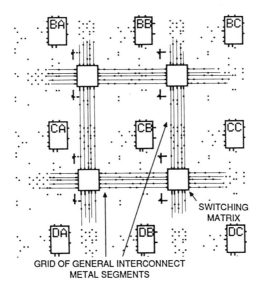

Figure 11-5 General-purpose interconnect grid. Logic cell array general-purpose interconnect is composed of a grid of metal segments which may be interconnected through switch matrices to form networks for CLB and I/O block inputs and outputs.

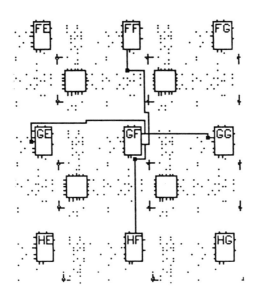

Figure 11-6 Direct interconnect grid. The .x and .y outputs of each CLB have single contact, direct access to inputs of adjacent CLBs.

The third level of interconnect resources is called long lines. These lines do not pass through the general-purpose switch matrices and are intended for signals that propagate over long distances on the device. These types of lines are shown in Figure 11-7. There are three vertical long lines running the height of the device between each column of CLBs and two horizontal long lines running the width of the device between each

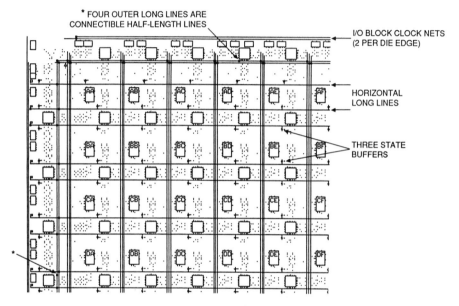

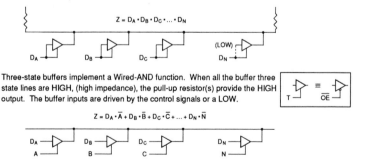

Figure 11-7 Long-line interconnect grid.

11.2 FPGA Architecture

two CLBs. There is some variation in the capabilities of long lines between different devices in the family members. However, any output from an I/O block or CLB can drive any long line through an isolation buffer. Two tristate buffers are located next to each CLB; their function is to drive the long lines. A buffer is not enabled onto any long line until it is actually used. Once a buffer is used, it must be tristated by the user to prevent logic contention when required. Since a "WIRED AND" function is performed by the tristate buffers, it is possible to implement multiplexers and other shared bus types of architectures. The long line feature is a powerful routing feature of the device and enables the designer to implement a large number of peripheral chip types of architectures in which several registers are accessed via a common data bus. A system designer must be careful with their use as the number of lines is limited and unacceptably long delays may be induced with poor placement.

The fourth routing level is a specialized set of long lines for a specific function. A special buffer is located on the device; its output drives a global net which inputs to all .k inputs to all CLBs. This permits use of an external clock input to drive the clock inputs of any selected CLB. Of course the .k input may also be programmed to take its input from any other routing resource. There is an alternate buffer which may be used to optionally drive the .k inputs to any CLB. This alternate allows for a second clock source for the CLB flip-flops.

11.2.3 XC3000 CLB Capabilities

Each configurable logic block (CLB) contains two flip-flops, a combinatorial logic function block, and a number of multiplexers. The CLB combinatorial function has five inputs (referred to as .a, .b, .c, .d, and .e) and two outputs (referred to as F and G) and is capable of generating any output which is a function of up to five inputs. Its inputs to each flip-flop can come from a variety of sources, including the F and G outputs from the combinatorial network, and direct data input (.di). The outputs from the CLB (.x and .y) can come from the flip-flops or from the F and G outputs of the combinatorial network. Feedback to the QX and QY inputs of the combinatorial network can come from the output of the flip-flops. There are also several control and clock lines. The enable clock (.ec) input allows the flip-flop D input to be sourced from either the flip-flop or from the combinatorial function or direct input. The flip-flop clock input (.k) is an invertable clock sent to both flip-flops. A reset (.rd) provides the capability of forcing the flip-flops to a cleared state. The reset is asynchronous.

The combinatorial functions are not implemented like a PLD using gates with programmable fuses or erasable fuses. All functions are implemented using a 32:1 look-up table technique so the implementation is more like that of a PROM.

11.3 Configuration

There are several techniques by which FPGAs are configured. The Xilinx parts are reconfigurable. The configuration or "metalization" of the part is lost each time power is removed from the part. Following a prescribed loading sequence, the programmable interconnect points are configured into the desired pattern when a power-up initialization sequence is performed. There are a variety of formats or modes in which the FPGA can be operated, supporting several different loading techniques. The development software provides either bit stream or byte parallel data files that can be loaded by a host CPU from disk, memory, or PROM files or the parts support a very simple serial bit stream loading feature using inexpensive 8-pin configuration PROMs. The largest device (XC3090) currently requires two configuration PROMs; larger configuration PROMs are scheduled for release in the near future. The device also supports a software reconfiguration sequence. Using this tool, several different configurations may be defined for the device. It is then possible to dynamically reconfigure the device while in use to fit different hardware requirements.

Actel devices use an "antifuse" technique to configure their devices. These are permanent configuration elements, so the "personality" of the device is not lost on power-down. The Actel devices also do not utilize the matrix switching technique but instead use a crossbar type of interconnect scheme. Using this method, most nets require only two antifuse links and result in a narrower range of net propagation delays. Net delays can also be reduced by the elimination of the matrix elements. Currently, the largest of the Actel devices has about 100,000 antifuses and routing is often 100% using the Actel supplied router.

11.4 FPGA Development Process

The process by which an FPGA is developed is conceptually simple but often complex in practice. There have been great strides made in the past few years to reduce entry costs and improve tools. All FPGAs are designed in a four-step process:

1. Design specification
2. Logic simulation

11.4 FPGA Development Process

3. Implementation
4. Target system test and debug

11.4.1 FPGA Design Specification

A Xilinx design can be specified using a variety of techniques. A common process involves schematic capture using FutureNet, CASE, Schema, PCAD, OrCAD, VIEWLogic, or other schematic capture packages. Each package comes with a standard library of hardware macro functions which can be called and connected to implement the desired function. There are also optional TTL libraries that can be used to implement higher level 7400 series functional blocks. In addition to schematic capture, PLDs may be embedded in the device by use of PLD macro library elements. The "fuse map" of the embedded PLD can be specified using standard PLD design software such as ABEL, CUPL, PLDesigner, and other software. This allows state machines to be easily hidden inside of the FPGA. The PLD file is then merged with the schematic input during postprocessing.

11.4.2 FPGA Simulation

Since it is not possible to "probe" internal connections or pins in an FPGA using conventional debugging techniques, the successful development of a solid design is heavily dependent upon simulation. A model of the device is constructed, given the net list of the device and characteristics of each element along with routing information. The model includes set and hold time requirements for flip-flops, clock rates, and propagation delays between every output and each input for all pins in each net. Simulated input signals are then used to excite the model and the outputs are calculated based on the model data base. Simulators build models with various degrees of sophistication. There is always uncertainty in the actual propagation delays. Some circuits may exhibit failure if the propagation delay is a minimum and set-up times are maximum, such as circuits with feedback, while others may exhibit failure if the propagation delays are maximum, such as rapidly clocked pipelined machines. Virtually all FPGA manufacturers support extensive simulation features and a number of second-source developers have the simulation and even development capability of FPGA manufacturers.

11.4.3 FPGA Implementation

There are two general methods by which a blank FPGA is programmed. The Xilinx method requires the programming to be performed each time the device is powered. It is done using one or more serial PROMs that are programmed from a JEDEC-compatible bit stream file created by the design and routing software. Devices may also be programmed by the byte loading method from a computer or other source. The Actel method uses antifuses and personality information is not lost on power-down. The antifuses make connections during the programming process, creating the required nets between logic functions.

11.4.4 FPGA Target System Test and Debug

There comes a day in the life of every FPGA design in which the first hardware device is actually plugged into the target system and testing begins. To ease the transition from mathematical nonentity to real working hardware, there are features built into most FPGAs hardware to permit gaining some understanding of the operation of the device. It would be useful to be able to attach logic analyzer probes to internal nets in an FPGA just as can be done in a board-level design. Fortunately, just such a testability feature has in fact been built into most FPGAs. These internal net signals can be monitored and downloaded to a host computer while the device is running in real time. The signals can be analyzed for failures like logic analyzer traces. The "probes" can be defined and moved using software commands.

11.5 Future Trends

Like PLDs, FPGAs are becoming increasingly popular and more common in design. Design tools are becoming increasingly more sophisticated and reliable, while costs are continuing to fall. It is now possible to design specialized (application-specific) processors, coprocessors, and peripherals in a matter of weeks with design tools and hardware in the $15,000 to $25,000 range, where as, only five to ten years ago, such designs would have required months or years, large staffs of engineers, and quite expensive development tools. This time can be likened to the golden age of systems integration. Increasingly, the division between the integrated circuits architect and the systems engineer is becoming indistinct. The combination of analog and digital circuits onto a common substrate has opened new horizons for data acquisition and control. Large and complex systems can be made more cheaply, quicker, and smaller with a fraction of the power consumption.

I recently completed the design of a 768-channel, 12-bit analog output board with a VMEbus interface and an 700 Mbit/s fiber-optic high-speed

11.5 Future Trends

load feature. Even eight years ago, to propose that such a design be implemented on a single board would have placed a design engineer in a place with well-padded walls. The project took only six months from conception to working model using techniques described in these two chapters.

System design, and especially bus design, is rapidly moving into a time in which system performance is going to make large strides forward again. Devices such as these will be required to meet the performance requirements of busses such as Fast bus, MicroChannel bus, and other architectures requiring massively parallel transfers and high speeds.

12

MULTIBUS I Design Example

12.0 Summary

This chapter and the next contain two simple examples of currently produced designs around MULTIBUS I and VMEbus. These two busses were selected due to the simplicity of MULTIBUS and the current popularity of VMEbus. Designs were selected that were not especially complex so that the principle of the design could be clearly illustrated.

12.1 MULTIBUS Digital-to-Analog Converter Board

The board used to illustrate a MULTIBUS design is manufactured by Datel and carries the name SineTrac ST-728. It supports either four or eight digital-to-analog (D/A) converter channels with a resolution of 12 bits. It supports 8- or 16-bit data transfers and has an absolute accuracy of 0.05% of full scale. This corresponds to an accuracy of two least significant bits. The output can be set to produce any of the following full-scale outputs:

1. ±5 VDC
2. ±10 VDC
3. 0 to +5 VDC
4. 0 to +10 VDC
5. 4 to 20 mA current loop

Each channel may be individually set to the desired output range or type. If current loop output is used, an external power supply must be provided that generates 18 to 30 VDC. Datel also has diagnostic software available for testing the board. Linearity error at 25°C is less than 0.5 LSB and over the maximum temperature range of 0° to 70°C is less than 1 LSB.

The board has been designed to be a memory-mapped peripheral occupying 16 contiguous memory locations. The base address may be set to any 16-byte boundary in address space. The memory usage of the ST-728 is shown in Figure 12-1. Note that the high address byte loads the upper 8 bits of the 12-bit D/A converter for any given channel, while the low address byte loads the lower 4 bits of the 12-bit D/A converter. Data bits D0 through D3 are "don't care" bits and are not used on the board. The board also has the optional capability of being ordered with a DC/DC converter that generates the ±15 V required for the linear circuitry on the board if the MULTIBUS implementation being used does not support these supply voltages. Figure 12-2 shows the board layout for the ST-728 with some key component locations shown.

DATA FORMAT

The ST-728 requires 12 bits of digital data, input from the host computer, for a single digital to analog (D/A) conversion. The chart below indicates how 8-bit and 16-bit CPU's format this data.

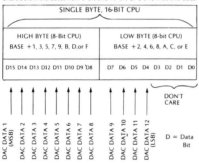

SELECTION OF 8-BIT OR 16-BIT CPU'S

The ST-728 board automatically changes to a 16-bit format when the BHEN/ line on the MULTIBUS goes to zero volts (pin 27 of the connector P1). A high input on BHEN/, consequently, sets the ST-728 for the 8-bit format.

ST-728 REGISTER ASSIGNMENTS

The following chart details the memory address assignments of the 16 memory locations the ST-728 occupies. Please note that when the ST-728 is used with 16-bit CPU's, every other (even-numbered) address location is used.

Note that 8-bit CPU's must transmit the 12 data bits in two bytes. The low byte contains the four least significant data bits: these are stored in a data register on the ST-728. The high byte contains the remaining 8 data bits. When the high byte is transmitted, all twelve bits of data — the 4 LSB's stored in a register and the 8 MSB's coming from the host computer — are loaded into the input of the selected DAC on the ST-728, and D/A conversion proceeds. Data transfer with a 16-bit CPU is somewhat simpler. All twelve data bits are transmitted in a single word. Data is loaded directly into the selected DAC, and a D/A conversion takes place.

ST-728 REGISTER ASSIGNMENTS

MEMORY ADDRESS (8-bit CPU's)	FUNCTION	REGISTER ASSIGNMENT	MEMORY ADDRESS (16-bit CPU's)
BASE +0	WRITE	Output LSB Byte for DAC 0 (Channel 0)	BASE +0
BASE +1	WRITE	Output MSB Byte for DAC 0 (Channel 0)	
BASE +2	WRITE	Output LSB Byte for DAC 1 (Channel 1)	BASE +2
BASE +3	WRITE	Output MSB Byte for DAC 1 (Channel 1)	
BASE +4	WRITE	Output LSB Byte for DAC 2 (Channel 2)	BASE +4
BASE +5	WRITE	Output MSB Byte for DAC 2 (Channel 2)	
BASE +6	WRITE	Output LSB Byte for DAC 3 (Channel 3)	BASE +6
BASE +7	WRITE	Output MSB Byte for DAC 3 (Channel 3)	
BASE +8	WRITE	Output LSB Byte for DAC 4 (Channel 4)	BASE +8
BASE +9	WRITE	Output MSB Byte for DAC 4 (Channel 4)	
BASE +A	WRITE	Output LSB Byte for DAC 5 (Channel 5)	BASE +A
BASE +B	WRITE	Output MSB Byte for DAC 5 (Channel 5)	

Figure 12-1 ST-728 memory map.

12.1 MULTIBUS Digital-to-Analog Converter Board

12.1.1 ST-728 Block Diagram

A block diagram of the ST-728 is shown in Figure 12-3. The most prominent features in the block diagram are the data registers, D/A converters (DAC) and I/O blocks. This four-block hierarchy is replicated four times on the ST-724, producing 4 output channels or eight times on the ST-728, producing eight output channels. The MULTIBUS interface consists of an address decoder, XACK delay generator, transfer type (8- to 16-bit) selector, data receivers with associated latch, and, optionally, the DC/DC converter.

The address decoder selects the channel to which data is being written and signals the XACK generator that a transfer is being performed. It also loads data into the least significant bit latch when transfers to the low-byte register are being performed. The data receivers buffer the data to be loaded into the selected D/A converter. The 8-/16-bit select circuitry examines the status of the BHEN/ line during a transfer to determine whether the transfer is a byte or word transfer. The appropriate data receivers are enabled based on the type of transfer being performed. Note that the board is a write-only board. Data loaded into the D/A converter registers may not be read back over MULTIBUS.

12.1.2 ST-728 Schematic Diagram

The schematic diagram of the ST-728 D/A board is shown in Figure 12-4. The heart of the output channels is the Datel DAC-681 12-bit D/A converter. The data sheet on the part is shown in Figure 12-5.

Figure 12-2 (a) ST723 D/A board. (b) Board layout. *(Figure continues.)*

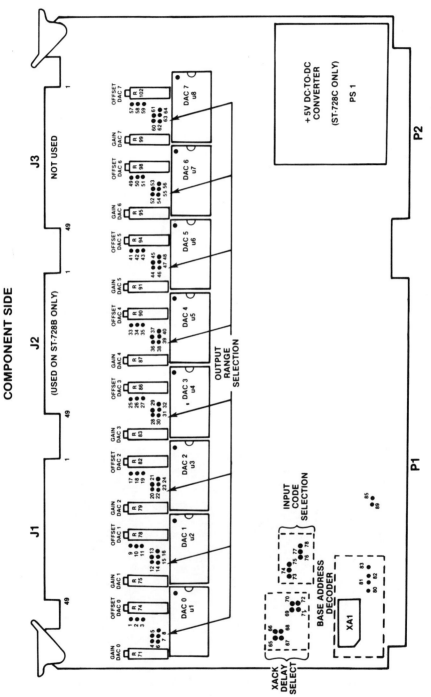

Figure 12-2 *(continued)*

12.1 MULTIBUS Digital-to-Analog Converter Board

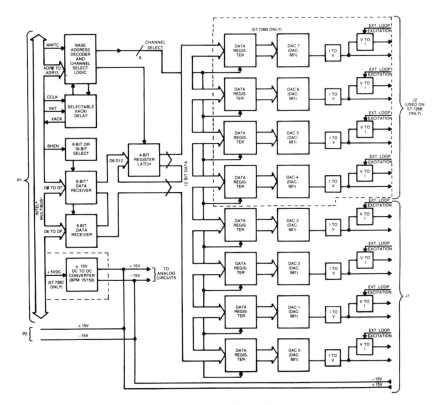

Figure 12-3 ST-728 block diagram.

Referring to Figure 12-4, if the bus master is performing an 8-bit transfer, U37 and U38 are enabled, causing 8-bit bus data to be placed on either DA1 through DA8 or DA9 through DA12, respectively. If the master is performing a 16-bit transfer, U36 and U38 are both enabled placing bus 16-bit data to be placed on DA1 through DA12. Note that, as is common in D/A converters, DA12 is the least significant and DA1 is the most significant bit.

U33, U34, and U35 are DM8136 identity comparators and are used for board address decoding. The upper 16 bits of the MULTIBUS 20-bit address space are compare with the address jumpered into jumper block XA-1. A board select signal is generated on pin 9 of U35 if the addresses match.

U31, a 74S138, selects which D/A converter (U9 through U24) will have data loaded after the transfer is complete. This is accomplished by decoding ADR1, ADR2, and ADR3 in conjunction with board select. The individual D/A select signals are given the net names STR0 through STR7.

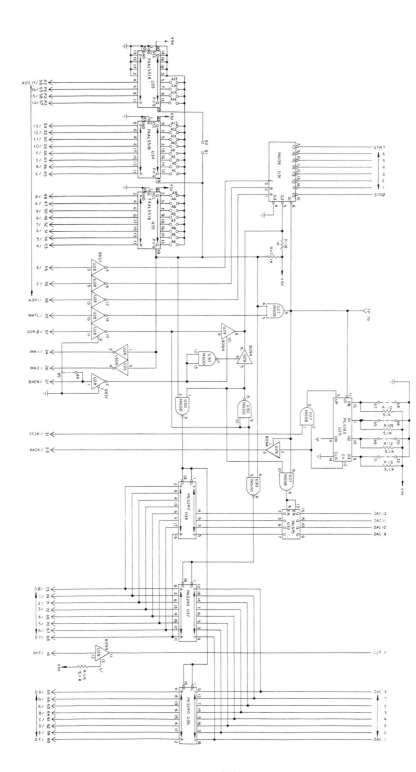

Figure 12-4 ST-723 schematic diagram. Reprinted by permission of DATEL, Inc. [Note: Unless otherwise specified, resistors are in ohms, ±5%, ½W; amplifiers are HA2-2505-2; capacitors are 0.1 MF; transistors are MPSU45; and diodes are IN4001.] *(Figure continues.)*

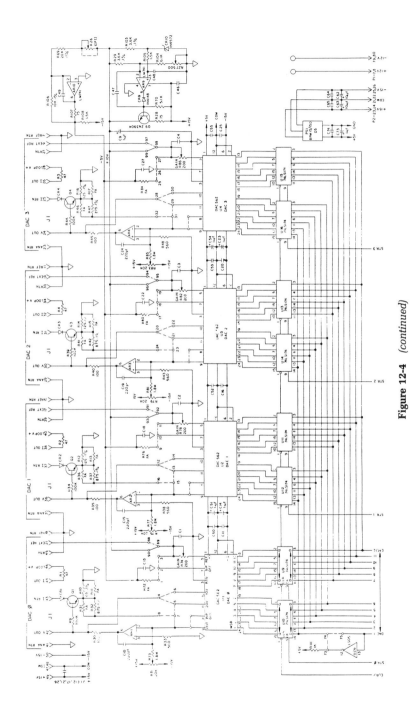

Figure 12-4 *(continued)*

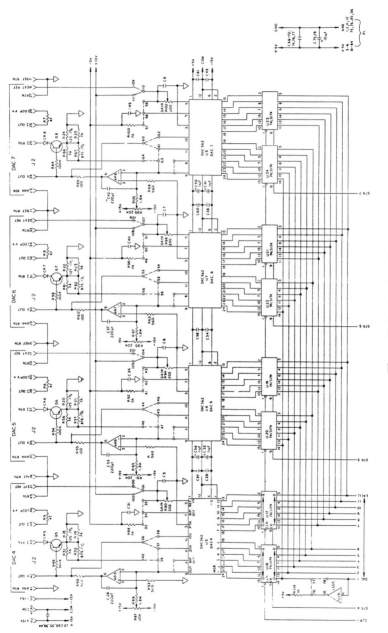

Figure 12-4 *(continued)*

12.1 MULTIBUS Digital-to-Analog Converter Board

U25, a 74LS193, is a binary counter used to generate a variable delay transfer acknowledge (XACK/). Jumpers can be changed on the D0 through D3 inputs to change the delay between the occurrence of the write command (MWTC/) and the generation of the XACK/ signal becoming active.

The D/A circuitry consists of eight identical sets of circuitry. D/A number seven reference designators will be used for the following dis-

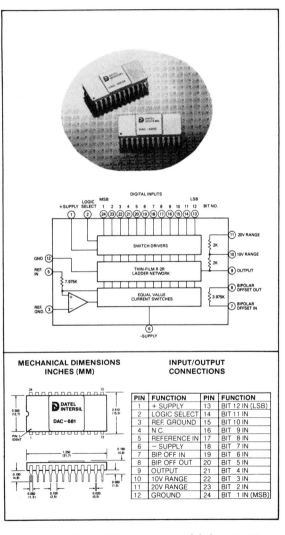

Figure 12.5 Monolithic high-performance 12-bit D/A converter model dac-681 *(Figure continues.)*

12. MULTIBUS I Design Example

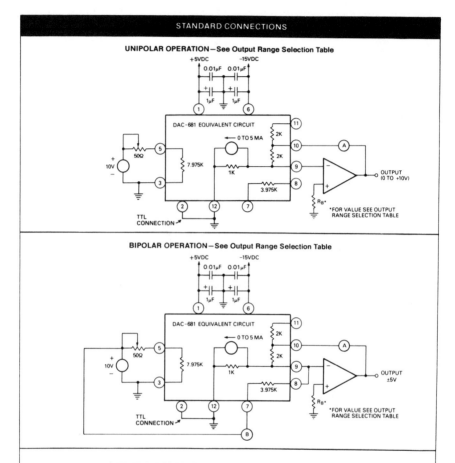

Figure 12-5 *(continued)*

12.1 MULTIBUS Digital-to-Analog Converter Board

SPECIFICATIONS, DAC-681
Typical at 25°C, +5V & 15V Supplies, +10V Reference unless otherwise noted.

	DAC-681C	DAC-681M
MAXIMUM RATINGS		
Positive Supply, pin 1	+20V	*
Negative Supply, pin 6	−20V	*
Reference Input, pin 5	±Supply	*
Reference Ground, pin 3	0V	*
Digital Inputs, pins 13-24	−1V to +12V	*
Logic Select Input, pin 2	−1V to +12V	*
Output, pin 9	+Supply, −5V	*
Resistors, pins 7, 8, 10, 11	±Supply	*
INPUTS		
Resolution	12 Bits	*
Coding, unipolar output	Straight Binary	*
Coding, bipolar output	Offset Binary	*
Input Logic Level, bit ON ("1")	+2.0 min. @ 100nA max	*
Input Logic Level, bit OFF ("1")	+0.8V max. @ −100µA max	*
Reference Input Voltage	+10V	*
Reference Input Resistance	8K	*
OUTPUTS		
Output Current, unipolar	0 to −5 mA	*
Output Current, bipolar	±2.5 mA	*
Output Voltage Ranges, unipolar	0 to +5V	*
	0 to +10V	*
Output Voltage Ranges, bipolar	±2.5V	*
	±5V	*
	±10V	*
Output Voltage Compliance	±1V	*
Output Resistance	1K	*
Output Capacitance	20 pF	*
PERFORMANCE		
Linearity Error, max	±½ LSB	±¼ LSB
Linearity Error Over Temp., max	±1 LSB	±1 LSB
Differential Linearity Error, max	±½ LSB	±¼ LSB
Monotonicity		Over Oper. Temp. Range
Gain Error, max.[2]	±0.25%	±0.10%
Unipolar Zero Error, max.[2]	±0.05%	*
Bipolar Offset Error, max.[2]	±0.25%	*
Gain Tempco, max.[3]	±10 ppm/°C	±10 ppm/°C
Zero Tempco, max.[3]	±2 ppm/°C	±2 ppm/°C
Bipolar Offset Tempco, max.[3]	±4 ppm/°C	±4 ppm/°C
Settling Time to ½ LSB[4]	300 nsec. typ., 400 nsec. max.	
Power Supply Sensitivity, max	±3.5 ppm of FSR/% Supply	*
Reference Slew Rate	6 mA/µsec.	*
Reference Bandwidth	10 MHz	*
POWER REQUIREMENT		
Rated Power Supply Voltage	+5 VDC −15 VDC	*
Positive Supply Range[5]	+5VDC±10%	*
Negative Supply Range	−15 VDC ±10%	*
Power Supply Quiescent Current, max	+15 mA, -40 mA	*
PHYSICAL-ENVIRONMENTAL		
Operating Temp. Range	0°C to +70°C	−55°C to +125°C
Storage Temp. Range	−65°C to +150°C	*
Package, Hermetically Sealed	24 pin ceramic	*

*Specifications same as first column

NOTES:
1. + Supply must be +5V ±5% for DAC-681C and +5V ±10% for DAC-681M. For operation with CMOS logic see Technical Note 1.
2. Adjustable to zero using external potentiometers. Specified error is for 24.9 ohm trim resistors and external op amp using internal feedback resistor.
3. Using external op amp and internal feedback and offset resistor. Zero Tempco and Bipolar Offset Tempco are in ppm/°C of FSR (Full Scale Range).
4. For full scale change: all bits ON-to-OFF, or all bits OFF-to-ON.
5. Maximum Positive Supply Voltage is +12V for high level logic only, i.e. when Pin 2 is tied to Pin 1. SEE Technical Note 1.

TECHNICAL NOTES

1. For TTL input logic, pin 2 should be connected to pin 12 and the + supply must be +5 VDC (±5% for DAC-681C and ±10% for DAC-681M). For CMOS input logic, connect pin 2 to pin 1 and use any + supply voltage from +4.75V to +12 VDC. CMOS threshold levels are then + Vs × 0.7 for bit ON and +Vs × 0.3 for bit OFF. Logic input current is the same as that specified for TTL.

2. Gain and bipolar offset errors are adjustable to zero by means of two 50 ohm trimming pots. The adjustment range is ±0.3% of FSR for gain and ±0.6% of FSR for bipolar offset. The unipolar zero error is adjustable to zero by means of the offset adjustment of the external output amplifier.

3. The output voltage compliance range of ±1V should not be exceeded or else accuracy will be affected. If a resistor load is driven instead of an op amp summing junction then the maximum resistor value is 200 ohms for unipolar operation and 400 ohms for bipolar operation.

4. Output settling time is specified for current output and is measured with a small current sampling resistor to ground (100 ohms). Voltage output settling time depends on the output operational amplifier used. Datel's AM-500 is recommended for about 500 nsec. settling and AM-452-2 is recommended for about 1.5 µsec. settling. Both should be used with a 3-20 pF variable compensating capacitor across the feedback resistor which should be adjusted for optimum settling time.

5. For best high speed performance, both power supplies should be bypassed with 1 µF electrolytics in parallel with 0.01 µF ceramic capacitors as close as possible to the ± supply pins.

6. The gain and bipolar offset temperature coefficients are specified with the internal feedback and offset resistors used in conjunction with an external operational amplifier. This is because these resistors track the R-2R ladder with temperature and therefore the tempco's do not depend on absolute resistor tempco. The tempco of the external +10V reference must also be included in the total converter tempco, however.

7. Because of the DAC-681 circuit which incorporates equally weighted current sources driving an R-2R ladder network, the turn ON and turn OFF times are virtually symmetrical, resulting in low output glitches compared with other DAC's. The major carry glitch typically has an amplitude of 14% of FSR. The time duration to 90% complete is typically 35 nsec.

8. The DAC-681 wideband output noise with all bits ON is typically 100 µV P-P over 0.1 Hz to 5 MHz.

ORDERING INFORMATION

Model	Temp. Range	Price (1-24)
DAC-681C	0 to 70°C	$ 24.50
DAC-681M	−55 to +125°C	$159.00

Trimming Potentiometer: TP50 – $3.50

THESE CONVERTERS ARE COVERED BY GSA CONTRACT.

Figure 12-5 *(continued)*

CALIBRATION AND APPLICATION

CODING TABLE — See Calibration Procedure

INPUT CODE	OUTPUT VOLTAGE RANGE				
	0 TO +5V	0 TO +10V	±2.5V	±5V	±10V
1111 1111 1111	+4.9988V	+9.9976V	+2.4988V	+4.9976V	+9.9951V
1100 0000 0000	+3.7500	+7.5000	+1.2500	+2.5000	+5.0000
1000 0000 0000	+2.5000	+5.0000	0.0000	0.0000	0.0000
0100 0000 0000	+1.2500	+2.5000	-1.2500	-2.5000	-5.0000
0000 0000 0001	+0.0012	+0.0024	-2.4988	-4.9976	-9.9951
0000 0000 0000	0.0000	0.0000	-2.5000	-5.0000	-10.0000

+10V REFERENCE CIRCUIT

Adjust R_4 for +10.000V output. For best stability R_1 & R_2 should track each other closely with temperature. R_4 should be a low tempco trimming pot or else a selected metal film trim resistor.

CALIBRATION PROCEDURE

1. Set all digital inputs LO. Adjust the output amplifier offset for 0 volts output.

2. Set all digital inputs HI. Adjust Gain trimming pot for an output of +FS-1LSB.
   FS-1LSB = +9.9976V for 0 to +10V range.
         = +4.9988V for 0 to +5V range.

1. Set all digital inputs LO. Adjust Bipolar Offset trimming pot for one of the following output voltages:
   -2.5V for ±2.5V range
   -5.0V for ±5V range
   -10.0V for ±10V range

2. Set bit 1 (MSB) input HI and all other digital inputs LO. Adjust Gain trimming pot for 0 volts output.

CIRCUIT FOR FAST VOLTAGE OUTPUT
($\approx$1.5 μSEC. SETTLING)

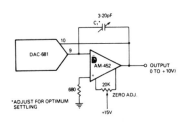

CIRCUIT FOR FAST VOLTAGE OUTPUT
($\approx$0.5 μSEC. SETTLING)

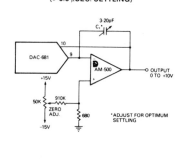

Figure 12-5 *(continued)*

12.1 MULTIBUS Digital-to-Analog Converter Board

cussion. U23 and U24 are a pair of 8-bit latches used to load the data bus data upon the occurrence of a register select signal STR7 on pin 9 of U24. U8, the D/A converter, generates an output current on pin 9 proportional to the binary input data. AR8 output operational amplifier functions as a transimpedance amplifier, converting the output current from the D/A converter (U8-9) to an output voltage on AR8-6. CR8, Q8, and associated passive components are used in the generation of the 4 to 20 mA current loop. Jumper positions are provided to enable the user to select the desired output voltage (or current loop) desired on a per channel basis.

AR10 and associated passive components generate a -5 VDC reference voltage for use by the D/A converter. AR9, Q9, CR9, and CR10 with associated passive components generate a $+10$ VDC reference voltage.

13

VMEbus Design Example

13.0 Summary

Like the preceding chapter, this chapter contains an example of a currently manufactured microcomputer board. The bus illustrated in this chapter is VMEbus. The example chosen for this chapter is manufactured by Dy4 Systems; it is their model DVME-503, 1 Mbyte memory board. It is a relatively simple example of a VME design. A simple example was selected purposely to illustrate the design process without confusing the illustration with unnecessary complexity.

13.1 VMEbus Memory Board

A photo of the DVME-503 board is shown in Figure 13-1. The board is a general-purpose memory board that can be configured for memory requirements from 64 kbytes to 1 Mbyte and accepts a large variety of RAM, ROM, or EEPROM devices. Battery backup for an installed RAM is also provided on the board. The board has 32 general-purpose sockets that can be configured to accept any of several devices. It will accept EPROMs ranging in size from 2 kbytes by 8 to 32 kbytes by 8, static RAM devices from 2 kbytes by 8 to 8 kbytes by 8, and EEPROM that is 2 kbytes by 8. A PLD is provided on the board so that the board can be configured to occupy continuous memory regardless of the size of the device installed on the board.

Additionally, the board is divided into four quadrants. Each quadrant may be independently configured for a different type of device. Using this

Figure 13-1 DY4 DVME-503 1-Mbyte memory board pictorial.

feature, the board may be used as a combination RAM, ROM, and EEPROM board. Provisions are also made on the board to accommodate slow-speed devices by jumper selections to slow down the access time of the board. The base address is user-selectable within a 24-bit address space. Both 16-bit (word) and 32-bit (long-word) data transfers are supported.

13.1.1 DVME-503 Block Diagram

A block diagram of the DVME-503 is shown in Figure 13-2. There are three major sections to this design. They are

1. VME interface
2. Memory array
3. Memory access control

13.1 VMEbus Memory Board

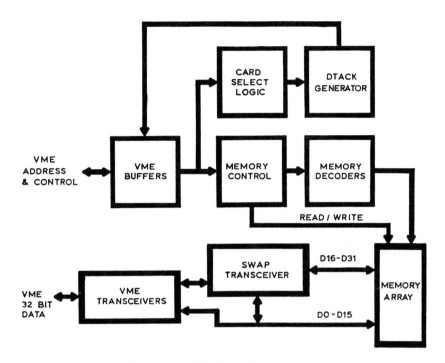

Figure 13-2 DVME-503 block diagram.

The VME interface consists of the VME address buffers and the 32-bit data transceivers. The buffers deliver address and control signals to on-board logic, while the data transceivers handle moving the data onto or off of the board during an access. The board base address can be placed on any 64k, 128k, 256k, 512k, or 1 Mbyte boundary, depending on the types of devices installed. The base address is determined by one of three methods. Jumper block JB36 may be used to set the base address, DY4 Systems can provide a PLD that factory-configures the board base address, or a PLD may be used in conjunction with a 3-bit code provided on the P2 connector. This code permits multiple boards to be installed in a system in which the memory maps are slot position-dependent. The memory selectors are 1:16 decoders that generate the memory device chip selects, depending on the address broadcast by the bus master.

The memory control section provides for control of the memory array by means of two PLDs (U42 and U43) and three jumper blocks, JB26, JB27, and JB28. JB27 is used for determining speed of access. Some devices, such as the 2186, require slow access times and the JB27 jumper is installed. Normal fast access is selected by removing the JB27 jumper. JB26 determines whether the board supports 16- or 32-bit transfers. JB28 determines one of four possible memory configuration options. Assuming the board is to be populated with devices that are all the same size, that size can be set to 2k, 8k, 16k, or 32k devices. If different size devices are to be used, a special PLD can be provided by the factory.

The memory array of 32, 28-pin sockets divided into four quadrants of 8 sockets each. The order of access is shown in Figure 13-3 for 16-bit access and Figure 13-4 for 32-bit access.

Numerical Order	Sockets Enabled (16 bit mode)	Even Byte A0=0	Odd Byte A0=1
16 (highest memory)	U44, U47	U44	U47
15	U38, U41	U38	U41
14	U45, U46	U45	U46
13	U39, U40	U39	U40
12	U34, U37	U34	U37
11	U26, U29	U26	U29
10	U35, U36	U35	U36
9	U27, U28	U27	U28
8	U25, U22	U22	U25
7	U15, U18	U15	U18
6	U24, U23	U23	U24
5	U17, U16	U16	U17
4	U14, U11	U11	U14
3	U4, U1	U1	U4
2	U13, U12	U12	U13
1 (lowest memory)	U2, U3	U2	U3

Figure 13-3 Socket order for 16-bit access.

Numerical Order	Sockets Enabled (32 bit mode)
8 (highest memory)	U38, U41, U44, U47
7	U39, U40, U45, U46
6	U26, U29, U34, U37
5	U27, U28, U35, U36
4	U15, U18, U22, U25
3	U17, U16, U24, U23
2	U1, U4, U14, U11
1 (lowest memory)	U2, U3, U13, U12

Figure 13-4 Socket order for 32-bit access.

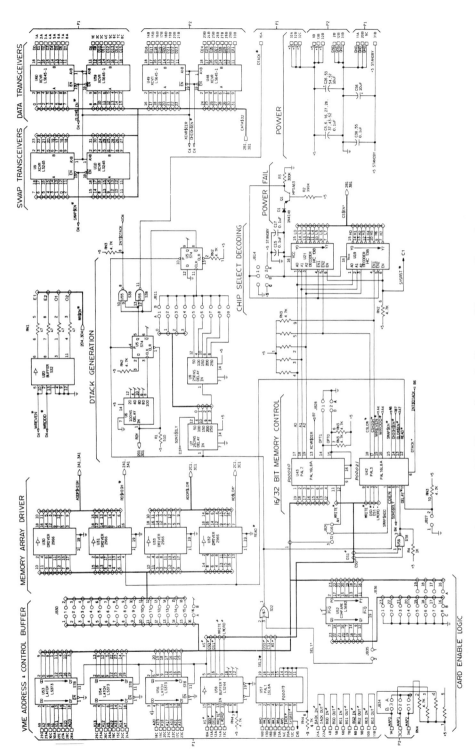

Figure 13-5 DVME-503 schematic diagram. *(Figure continues.)*

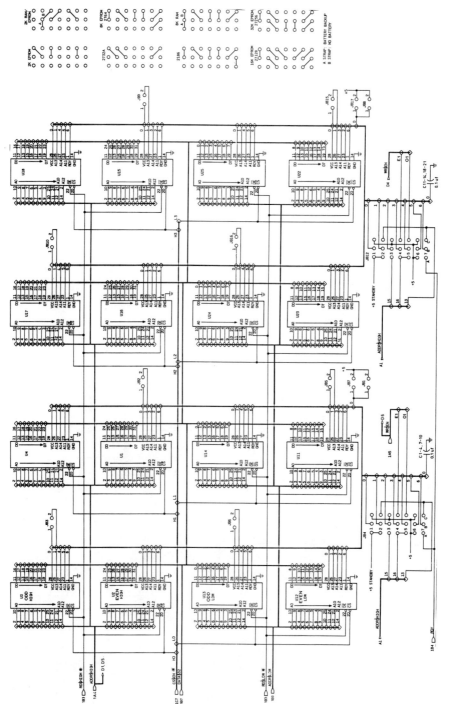

Figure 13-5 *(continued)*.

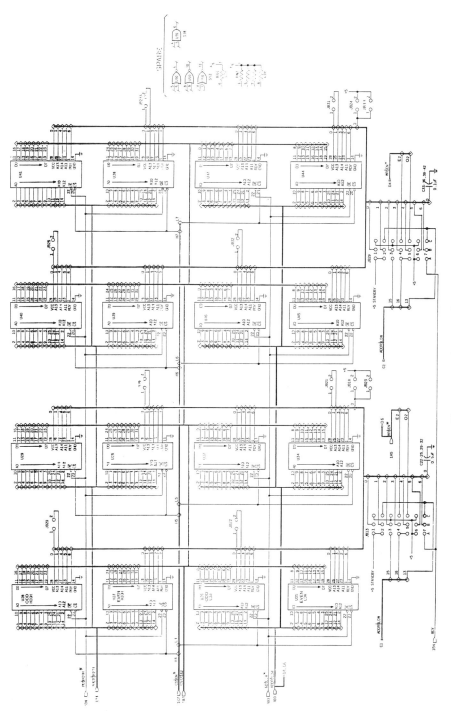

Figure 13-5 (continued).

13.1.2 DVME-503 Schematic Diagram

The schematic diagram of the memory board is shown in Figure 13-5. The VME address buffers are devices U53, U54, U56, and U58. Additionally, address modifier decoding is performed by PLD, U57. Since the board has no provisions to be a bus master or generate interrupts, IACK and BGn signals are bussed through the board. PLD U51 provides for decoding a 3-bit code supplied by the P2 user-defined pins. Buffers U30, U31, U32, and U33 are used to buffer address information to the memory array. U5 through U8 are used to set the DTACK generation delay during a board access. Jumper block JB11 is used to determine the time delay to when the handshake occurs. PLDs U42 and U43 perform the function of memory access control while U19 and U21 provide for memory array chip select decoding. U48, U49, U59, and U60 are data bus transceivers moving data from the local data bus to the VMEbus or conversely. U9 and U10 move data from the upper 16 bits of the local data bus to the lower 16 bits during a 16-bit access to the upper 16 bits of the memory array. Pages 2 and 3 of the schematic are simply the 32 memory array devices.

Index

ABEL source code
  programmable logic devices, 182
  programming example, 193–202
Ack cycle, NuBus, 121–122
Address bus, signal groups and uses, 5
Address/data group, MULTIBUS II line descriptions, 67
Address lines
  MULTIBUS system bus, 42–43, 45
  MULTIBUS II, 73
    timing requirements, 76–77
  STD bus, 156
  VMEbus, 88, 90
Address map, programmable logic devices, 188–190
Address only cycle VMEbus, 87, 94–95
Analog models, transmission-line concepts, 6
Analog-to-digital converters (ADCs), programmable logic devices, 188–190
AND gates, programmable logic devices, 174
ANSI/IEEE Std 1101–1987 standard, 60
APLUS source code
  programmable logic devices, 182
  programming example, 190–193
ARB5/-ARB0/, MULTIBUS II line descriptions, 71–72
Arbiter, VMEbus, 87
Arbitration algorithms, VMEbus, 103, 105–106

Arbitration lines
  MULTIBUS II, 65, 67–68
  multiple masters, 52–54
  NuBus, 120–121, 128–130
  STD bus, 158–159
  VMEbus, 103–104
Arbitration priority reset timing, 74
Arbitration process, NuBus arbitration, 129–131
Arbitration signal group, MULTIBUS II, 71–72
Architecture, field-programmable gate arrays, 204–212
Assemblers, software/hardware interactions, 28–30
AT board dimensions, PC/XT/AT bus, 137, 139
AT bus lines, 143–146
Attention cycle, NuBus, 121
Attention operations, NuBus arbitration, 128–129
Availability, of busses, 3

Backplane transmission lines, 11–12
Block operations
  MULTIBUS II line descriptions, 72
  NuBus, 124–128
  VMEbus, 96, 100–101
Block read cycle, VMEbus, 86–87
Block write cycle, VMEbus, 86–87

239

Board spacing, STD bus, 154–155
Board thickness, NuBus, 119
Board transmission lines, 11–12
Boolean equation review, programmable logic devices, 166–168
BREQ/, MULTIBUS II, 72
Broadcast message, MULTIBUS II, 75–76
Bubble chart, programmable logic devices, 169
Buffered driver connection scheme, 15–16
Bus
 basic concepts, 1–2
 comparisons
  data transfers, 19–20
  functions, 18–19
  interrupts, 21, 23
  limitations, 17–18
  multiple masters, 22–23
  utility functions, 23, 25
 defined, 2
 design benefits and drawbacks, 2–4
 signal groups and uses, 4–6
 transmission-line concepts, 6–16
 terminations and reflections, 10–16
Bus busy (BUSY/), MULTIBUS control bus, 47
Bus clock (BCLK/), MULTIBUS control bus, 46–47
Bus cycles
 PC/XT/AT bus, 146–152
 VMEbus, 86–87
BUSERR/ signals, MULTIBUS II,
 line descriptions, 71
 timing requirements, 78
Bus exchange
 MULTIBUS system, 55–56
 VMEbus, 106–108
Bus inhibit operations, MULTIBUS data transfers, 50–52
Bus interaction, POP instructions, 32–35
Bus interface device (PLX464) family, 180–182
Bus lines
 NuBus, 119–121
 PC/XT/AT bus, 139–146
 STD bus, 156–159
Bus priority in (BPRN/), MULTIBUS control bus, 47
Bus priority out (BPRO/), MULTIBUS control bus, 47

Bus request (BREQ/), MULTIBUS control bus, 46–47
Bus timer, VMEbus, 87
Bus-vectored interrupts
 MULTIBUS system, 56–57
 STD bus, 153
Byte high enable (BHEN/), MULTIBUS control bus, 48

Capacitance, transmission lines, 14–16
Card cages, MULTIBUS II, 61
Carnaugh map form, programmable logic devices, 166–168
CCLK/, MULTIBUS II line descriptions, 72
Central control group, MULTIBUS II, 72–73
Central processing units (CPU)
 iLBX bus, 40
 PC/XT/AT bus, 135–136
Central services module (CSM)
 MULTIBUS II, 65, 67–68
  line descriptions, 71
  timing requirements, 73–74
Ceramic-based board machines, 8–9
Clad density, transmission-line concepts, 8
Clad run, transmission-line concepts, 7
Code segment (CS), PC/XT/AT bus, 151–152
Command lines, MULTIBUS data transfers, 50
Common bus request (CBRQ/), MULTIBUS control bus, 47
Compatibility, STD bus data transfer, 159–160
Configurable logic blocks (CLBs), 205–211
Constant clock (CCLK/), MULTIBUS control bus, 46
Control bus
 MULTIBUS system bus, 46–49
 signal groups and uses, 5
Control lines, VMEbus, 90–92
Costs, bus specifications, 3
CRBQ/ line, parallel prioritization, 52–54

Daisy-chain driver connection schematic, 15–16
Data bus signals
 NuBus, 120
 signal groups and uses, 5

Index

Data lines
  MULTIBUS system bus, 42, 44
  STD bus, 156
  VMEbus, 90–92
Data transfer bus, 18–20
  VMEbus, 82, 90–101, 103–108
Data transfer lines
  MULTIBUS control bus, 47–48
  STD bus, 157, 159–163
Data transfers
  MULTIBUS, 49–52
    bus inhibit operations, 50–52
    data read operations, 49–51
    data write operations, 49–50
  NuBus, 122–128
Data write operation, MULTIBUS data transfers, 49–50
DCLOW/, MULTIBUS II, 73
Design packaging
  bus design and, 4
  field-programmable gate arrays, 213
Dielectric constant, transmission line concepts, 9
Digital-to-analog converters (DAC)
  MULTIBUS I design example, 217–229
  programmable logic devices, 188–190
Direct memory access, iSBX bus, 39
Disk operating system (DOS), PC/XT/AT bus, 136
DispTimeDate procedure, 32–33
Distributed handler systems, VMEbus, 108
DMA operations, PC/XT/AT bus, 148–151
  read operation, 148–149
  write operation, 150–151
Double height board, VMEbus, 83–84
Driver characteristics, NuBus arbitration, 132–134
Dual-mode boards, MULTIBUS II, 63
DVME-503, VMEbus design example, 231–238

Electrical specifications, MULTIBUS II, 80
End of transfer (EOT), MULTIBUS II, 63, 65–66
Epoxy-based board machines, 8–9
EPROMs, programmable logic devices, 173
Equation reduction, programmable logic devices, 168
Exception operations, MULTIBUS II, 65

line descriptions, 71
timing requirements, 77–78, 80
"Extended" memory, PC/XT/AT bus, 137

Field-programmable gate arrays
  architecture, 204–212
    interconnection resources, 206, 208–211
    I/O characteristics, 205–208
    XC3000 CLB capabilities, 211–212
  configuration, 212
  design specification, 213
  development process, 212–214
  future trends, 214–215
  implementation, 213–214
  overview, 203–204
  simulation, 213
  speed, 204
  target system test and debug, 214
Field programmable logic sequencer (FPLS) (82S105), 176–177
Form factor board, NuBus, 116, 118
Full-sized boards, PC/XT/AT bus, 137–138
Functional modules, VMEbus, 87–89
Future bus (P896) standard. *See* MULTIBUS II

Generalized block diagram, 168–170
Geographic addressing, NuBus arbitration, 130
GetDate procedure, 30–31
GetTime procedure, 28–30

Handshaking protocol
  MULTIBUS system bus, 41
  programmable logic devices, two-wire handshake, 188
Hardware architecture, bus limitations and, 17–18
High-level languages, software/hardware interactions, 30, 32

IACK daisy-chain driver operation
  VMEbus, 87
    priority interrupts, 112–113

IACKIN*/IACKOUT* daisy chain, VMEbus, 109–110
  priority interrupts, 109–110
IACK* line, VMEbus, 111
IDLACH/, MULTIBUS II line descriptions, 72–73
IEEE-796 standard, MULTIBUS system bus, 41
iLBX bus, 39–40
Impedance
  microstrip backplane transmission lines, 10
  transmission-line concepts, 7–8
    source and line characteristic impedances, 14–15
Inhibit (INH1/ and INH2/), MULTIBUS control bus, 48
Initialize (INIT/), MULTIBUS control bus, 46
Input/output space, MULTIBUS II timing requirements, 77
Interconnection resources, field-programmable gate arrays, 206, 208–211
Interconnect space, MULTIBUS II timing requirements, 76–77
Interrupt acknowledge cycle, VMEbus, 87
Interrupts
  bus comparisons and, 18, 21, 23
  MULTIBUS system, 48–49, 56–57
  PC/XT/AT bus, 151–152
  STD bus, 157
  two-cycle and three-cycle interrupts, 57
  VMEbus, 87
    priority interrupts, 111–113
I/O block diagram, programmable logic devices, 184, 186
I/O characteristics
  field-programmable gate arrays, 205–208
  read operations, PC/XT/AT bus, 146–147
  write operations, PC/XT/AT bus, 147–148
IRQ line, VMEbus, priority interrupts, 109, 111–113
IRQn line, VMEbus, 111
iSBX bus, 39

JEDEC file, programmable logic devices, 183

Kirchhoff's current and voltage laws, 12–13

Large-scale device (EP1800), programmable logic devices, 177–180
Line characteristics, NuBus arbitration, 132–134
Line descriptions, MULTIBUS II, 67, 69–73
  address/data group, 67
  arbitration signal group, 71–72
  central control group, 72–73
  exception operation group, 71
  system control group, 69–71
Location monitor, VMEbus, 87
Lock (LOCK/), MULTIBUS control bus, 46
LOGIC CAPS, programmable logic devices, 182–183
Long-line interconnect grids, field-programmable gate arrays, 210–211
Long-word (quad-byte) read and write, VMEbus, 96, 99

Master module, VMEbus, 87
Master/slave bus transfer, MULTIBUS system bus, 41
Mechanical features, bus comparisons and, 24–25
Medium-scale devices (20RS10), programmable logic devices, 174–175
Memory board, VMEbus design example, 231–238
Memory read operations, PC/XT/AT bus, 146–147
Memory space
  MULTIBUS II, timing requirements, 77
  write operations, PC/XT/AT bus, 147–148
Message space, MULTIBUS II timing requirements, 77
Metastability, programmable logic devices, 183–185
Microstrip transmission line, 8–10
MOV DX, REGSEC timing diagram, MULTIBUS interaction and, 34
MULTIBUS boards, assemblers for, 28
MULTIBUS I system
  address lines (A0/ through A23/), 42–43, 45

Index

control bus, 46–49
  data transfer lines, 47–48
  interrupt lines, 48–49
  utility and bus arbitration lines, 46–47
data transfers, 42, 44, 49–52
  bus inhibit operations, 50–52
  read operations, 49–51
  write operations, 49–50
design example
  board layout, 220
  digital-to-analog converter board, 217–229
  monolithic high-performance model, 225–229 ST-728
  block diagram, 219, 221 ST-728
  schematic diagram, 219, 221–229
iLBX bus, 39–40
IN AL, DX timing diagram, 34–35
interrupt lines, 48–49, 56–57
  bus and non-bus vectored interrupts, 56–57
iSBX bus, 39
multichannel I/O bus, 40
multiple masters, 52–56
  bus arbitration, 52–54
  bus exchange, 55–56
outline and dimensions, 37–38
overview of, 37–39
PI connector pin assignment, 43, 45
power logic levels, 41–43
system bus, 41–49
termination and drive requirements, 44
summary, 217
MULTIBUS II
  board dimensions, 61–62
  electrical specifications, 80
  features and capabilities, 60
  line descriptions, 67, 69–73
    address/data group, 67
    arbitration signal group, 71–72
    central control group, 72–73
    exception operation group, 71
    system control group, 69–71
  mechanical specifications, 60–63
  operations, 63, 65–67
  overview, 60
  signal groups and uses, 5
  timing requirements, 73–80
    exception operations, 77–78, 80
    reset operations, 73–74

    transfer operations, 73, 75–76
    address space, 76–77
    broadcast messages, 75–76
Multichannel I/O bus, MULTIBUS system, 40
Multilayer boards, transmission-line concepts, 8
Multiple masters
  bus comparisons and, 18, 22–23
  MULTIBUS system, 52–56
Multiple ("star-type") connection schematic, 14–15

Non-bus-vectored interrupts, MULTIBUS system, 57
NuBus
  bus lines, 119–121, 128–130
    arbitration bus lines, 120–121, 129–131
    attention operations, 128–129
    data bus signals, 120
    power, 119
    utility signals, 119–120
  cycles and transactions, 121–128
    ack cycle, 121–122
    attention cycle, 121
    block operations, 124–128
    data transfers, 122–128
    read transfers, 123–124
    start cycle, 121
    write transfers, 124–125
  electrical characteristics, 132–134
  geographic addressing, 130
  mechanical specifications, 116–119
    PC board, 117–119
    triple-height board, 116–117
  overview, 115–116
  pin assignment, 133–134
  power supply voltages, 133–134
  utility functions, 130, 132

OR gates, programmable logic devices, 174
Output enable (OE) input, 177
Outputs, programmable logic devices, 169

PALASM, programmable logic devices, 182
Parallel prioritization, bus arbitration, 53–54

Parallel system bus (PSB), MULTIBUS II, 60
Parity checking, NuBus data transfer, 123
Pascal source, example, 31–32
PC/XT/AT bus
  bus cycles, 146–152
    DMA operations, 148–151
      read operation, 148
      write operation, 151
    interrupt operations, 151–152
    memory or I/O read operations, 146–147
    memory or I/O write operations, 147–148
  bus lines, 139–146
    AT-specific lines, 143–146
    pin assignments, 139–143
  features and capabilities, 136–137
  mechanical specifications, 137–139
  overview, 135–136
Pin assignment
  AT bus lines, 143–146
  MULTIBUS II, 64
  NuBus arbitration, 134
  PC/XT/AT bus, 140–141
  PI connector, MULTIBUS system bus, 43, 45
  VMEbus, 85
PLX technology, programmable logic devices, 184
"Pointers," bus interaction and, 32–35
POP instructions, assembly language source module and, 30
Portable code, high-level languages and, 30, 32
Power levels
  bus design and, 4
  logic levels, MULTIBUS system bus, 43
  MULTIBUS system bus, 41–43
  NuBus, 119
  STD bus, 156
  signal groups and uses, 5–6
  supply voltages, NuBus arbitration, 133–134
Power monitor, VMEbus, 87
Preset input (PR), programmable logic devices, 177
Printed circuit boards, transmission-line concepts, 8

Prioritized (PRI) arbiter, VMEbus, 105–106
Priority interrupt bus, VMEbus, 82, 108–113
Program counter (PC), PC/XT/AT bus, 151–152
Programmable array logic (PAL), 171–173
Programmable interrupt controller (PIC), 151–152
Programmable logic devices
  architecture, 171–182
    bus interface device (PLX464) family), 180–182
    field programmable logic sequencer (FPLS) (82S105), 176–177
    large-scale device (EP1800), 177–180
    medium-scale devices (20RS10), 174–175
    small-scale devices (16L8), 171–174
  Boolean equation review, 166–168
  overview, 165–166
  programming, 182–183
    examples of, 184, 186–201
      ABEL code, 193–202
      APLUS source code, 190–193
    state machine review, 168–170
PROM files, field-programmable gate arrays, 212
PROT/, MULTIBUS II line descriptions, 73

Random access memory (RAM)
  PC/XT/AT bus, 136
  VMEbus, unaligned transfers, 93–94
Read-modify-write (RMW) operations
  MULTIBUS control bus, 46
  VMEbus, 86–87, 101–102
Read operations
  MULTIBUS data transfers, 49–51
  NuBus, 123–124
  STD bus data transfer, 160–163
  VMEbus, 86–87
Real-time clock (RTC), assembly language source module and, 28–29
Reduced-size board, PC/XT/AT bus, 137–138
REGSEC variable, bus interaction and, 32–35
Replying agents, MULTIBUS II, 63
Reply phase function, MULTIBUS II, 69

Index

Requesting agents
  MULTIBUS II, 63
  VMEbus, 87
RESET/, MULTIBUS II line descriptions, 73
Reset operations, MULTIBUS II timing requirements, 73-74
Resistance, transmission lines, 14-16
ROM, PC/XT/AT bus, 136
Round robin select (RRS) arbiter, 106
RSTNC/, MULTIBUS II line descriptions, 73

S and R equations, programmable logic devices, 177
SC0, MULTIBUS II control signals, 71
SC1 MULTIBUS II control signals, 71
SC4/-SC5/, MULTIBUS II control signals, 70-71
SC7/-SC5/ control signals, 70
SC8/ control signals, 70
SC9/-SC0/ control signals, 69
Seconds variable, bus interaction and, 32-35
Serial prioritization, bus arbitration, 52-53
74138 decoder, bus arbitration, 53-54
74148 priority encoder, bus arbitration, 53-54
Signal lines
  bus design and, 4-6
  STD bus, 155-159
Signal voltage tolerances, STD bus data transfer, 163
Simulation, field-programmable gate arrays, 213
Single-board computers
  bus design and, 3
  iSBX bus, 39
Single-byte read and write, VMEbus, 95-97
Single handler systems, VMEbus priority interrupts, 108
Single-level (SGL) arbitration schemes, VMEbus, 105
Slave module, VMEbus, 87
Small-scale device (16L8), programmable logic devices, 171-174
"Sneak pulses," programmable logic devices, 169
Software, PC/XT/AT bus, 135-136

Software/hardware interactions, 27-35
  assemblers, 28-30
  bus interaction, 32-36
  high-level languages, 30, 32
Speed
  bus design and, 3-4
  field-programmable gate arrays, 204
Star-fed transmission lines, 14-16
Start cycle, NuBus, 121
State machine diagram, programmable logic devices, 184, 187
State machine review, programmable logic devices, 168-170
STATUS/ID information, VMEbus priority interrupts, 109, 111-113
STD bus
  data transfer operations, 159-163
  compatibility, 159-160
  read/write operations, 160-163
  signal voltage tolerances, 163
  mechanical outline, 154-155
  overview, 153-154
  signal groups and uses, 4-5
  signal lines, 155-159
    address and data lines, 156
    bus control lines, 156-159
    arbitration lines, 158-159
    data transfer lines, 157
    interrupt control lines, 157
    STD timing lines, 157-158
    utility lines, 159
    power and auxiliary power, 156
Stripline transmission line, 8-9
System clock driver, VMEbus, 87
System control group, MULTIBUS II line descriptions, 69-71
System control (SC) lines, MULTIBUS II timing requirements, 73

Target system test and debug, field-programmable gate arrays, 214
Termination and drive requirements, MULTIBUS system bus, 43-44
Three-cycle interrupts, MULTIBUS system, 57
TIMEOUT/, MULTIBUS II
  line descriptions, 71
  timing requirements, 78

Timing lines, STD bus, 157–158
Timing requirements, MULTIBUS II, 73–80
   exception operations, 77–78, 80
   reset operations, 73–74
   transfer operations, 73, 75–77
Transfer acknowledge (XACK/)
   MULTIBUS control bus, 48
   MULTIBUS data transfers, 50
Transfer completion status, NuBus, 122–123
Transfer operations
   MULTIBUS II, 63, 65–66
      timing requirements, 73, 75–76
      protocol, 41
Transition conditions, programmable logic devices, 197, 199
Transmission-line concepts
   bus design, 6–16
   terminations and reflections, 10–16
Triple-height boards, NuBus, 116–117
Two-cycle interrupts, MULTIBUS system, 57
Two-layer boards, transmission-line concepts, 8

Unaligned transfers, VMEbus, 93–94
User-defined (UD) pins, VMEbus, 85
Utility functions
   bus comparisons and, 18, 23, 25
   MULTIBUS control bus, 46–49
   NuBus arbitration, 119–120, 130, 132
   STD bus, 159
   VMEbus, 82

VMEbus
   basic features, 82
   board dimensions, 83–84
   cycles, 86–87
   data transfer arbitration bus, 88, 90–101, 103–108
      A01–A31 address lines, 88
      address-only broadcast, 94–95
      AM0–AM5 address lines, 88, 90
      arbitration algorithms, 103, 105–106
      arbitration lines, 103–104
      AS*, 91
      BERR*, 91
      bus exchange, 106–108
      D00-D31, 91
      DS0* & DS1*, 90–92
      DTACK*, 91
      LWORD*, 91
      read-modify-write operations, 101–102
      single-byte read and write, 95–97
      unaligned transfers, 93–94
      WRITE*, 91
   design example, 4
      memory board, 231–238
         block diagram, 232–233, 238
         schematic diagram, 235–238
      summary, 231
   functional modules, 87–89
   mechanical specifications, 82–85
   overview, 81–82
   pin assignments, 85
   priority interrupt bus, 108–113
      IACK daisy-chain driver operation, 112–113
      IACKIN*/IACKOUT* daisy chain, 109–110
      interrupt process, 109, 111–113
      programmable logic devices, state machine diagram, 184, 187
   signal groups and uses, 4–5
VME controller, programmable logic devices, 188–190

Waveforms, transmission-line, 10–12
WIRED AND function, field-programmable gate arrays, 211
Wiring techniques, transmission-line concepts, 8
Word (double-byte) read and write, VMEbus, 96, 98
Write cycle, VMEbus, 86–87
Write operations, STD bus data transfer, 160–163
Write transfers, NuBus, 124–125

XC3000 CLB capabilities, field-programmable gate arrays, 211–212
Xilinx architecture
   field-programmable gate arrays, 205–211
   programmable logic devices, 183–185
XOR gate, programmable logic devices, 174–175
XT board dimensions, PC/XT/AT bus, 137–138